MECHANISMS
OF MOLECULAR
EVOLUTION

MECHANISMS OF MOLECULAR EVOLUTION

Introduction to Molecular Paleopopulation Biology

Edited by Naoyuki Takahata
National Institute of Genetics, Mishima, Japan

and Andrew G. Clark
The Pennsylvania State University,
University Park, Pennsylvania, U.S.A.

JAPAN SCIENTIFIC SOCIETIES PRESS
Tokyo

SINAUER ASSOCIATES, INC.
Sunderland, Massachusetts

Supported in part by the Ministry of Education, Science and Culture under Grant-in-Aid for Publication of Scientific Research Result.

Published jointly by:
JAPAN SCIENTIFIC SOCIETIES PRESS
2-10 Hongo, 6-chome, Bunkyo-ku, Tokyo 113, Japan
ISBN 4-7622-6718-X

and

SINAUER ASSOCIATES, INC.
Sunderland, Massachusetts 01375, U.S.A.
ISBN 0-87893-825-7

Distributed in all areas outside Japan and Asia between Pakistan and Korea by Sinauer Associates, Inc.

Printed in Japan

Preface

This volume comprises 13 papers selected from the 17th Taniguchi International Symposium on Biophysics held November 11-15, 1991 in Mishima, Japan. The main topic was population biology, in light of the wealth of new data at the DNA level, and emphasis was placed on population aspects of molecular data rather than simply on evolution of the individual molecules. Hence the title: *Population Paleo-Genetics* or *Molecular Paleo-Population Biology*. The basic theme was similar to what the late Professor Hitoshi Kihara expressed as early as 1947: *The history of the earth is recorded in the layers of its crust; the history of all organisms is inscribed in the chromosomes*. With the introduction of DNA technology into evolutionary biology and recent advances in population genetics, however, we are in a position to better understand the history of organisms as well as the evolutionary mechanisms. The aim of this book is to demonstrate excitement stemming from recent developments in this field.

Much theoretical work has been done on genealogical processes arising at genetic loci in finite populations, and has proved extremely useful for understanding changes of genetic information at the DNA level as well as inferring the evolutionary history of organisms. Takahata reviews the allelic genealogy under balancing selection in a finite population and derives approximate formulas concerning gain and loss of alleles when population size changes. Applying these formulas to alleles at the major histocompatibility complex loci (*Mhc*), he quantitatively examines the history of modern humans and three hypotheses for the origin of our species. Hudson describes methods of generating one- and two-locus gene genealogies to illustrate their use and power in examining properties of test statistics under the assumption of neutrality. Specifically, he explores the sampling properties of the joint distribution of allele frequencies and linkage disequilibrium, and proposes a new estimator of the recombination parameter. Tajima presents his own work on measuring DNA polymorphism. Of particular interest are the statistical properties of the number of pairwise nucleotide differences and the number that are related to segregating sites in a sample. These are examined with and without selection, migration, or changes in population size. Golding points out that distinct patterns of genetic changes occur in phylogenies when selection acts on DNA sequences, and uses such patterns to search for the influence of selection *via* a maximum likelihood approach. He shows that the algorithm can detect even very weak selection.

There is growing evidence of natural selection acting at the DNA level. Some examples of this are well known—self-incompatibility in plants, *Mhc* in vertebrates, and immunogenic proteins in parasites. Clark provides a comprehensive review of theory and molecular data of S-protein polymorphism involved in gametophytic self-incompatibility. Like *Mhc* genes, S-protein alleles show the *trans*-specific mode of evolution and the polymorphism is often shared by distantly related species. He shows the primary importance of mutation and selection, but no evidence for intragenic recombination generating new alleles. Analysis of DNA sequences from the circumsporozoite protein and merozoite surface antigen-1 loci of *Plasmodium falciparum* leads Hughes to the conclusion that polymorphisms at these loci are maintained by balancing selection favoring evasion of host immune recognition. He suggests that *P. falciparum* was transferred to humans from another vertebrate host long ago and the immunogenic proteins have coevolved with the host's immune system. Satta makes an extensive analysis of DNA sequences of *Mhc* alleles. She demonstrates that balancing selection acting on the nonsynonymous changes in the peptide-binding region is much stronger than previously thought, and therefore the rate of substitution is much faster than the synonymous one. Estimates of the synonymous substitution rate, the degree of selective constraint and the number of alleles at an *Mhc* locus, and clear instances of intragenic recombination or gene conversion in Class I Mhc molecules are presented. Based on sequence information of the *t*-complex polypeptide-1 (*Tcp*-1) gene, Morita *et al.* show that *t*-haplotypes emerged much earlier than the genus *Mus musculus* diverged. However, since the *Tcp*-1 genes of *t*-haplotypes sampled from different *M. musculus* subspecies are identical, they conclude there has been a recent introgression of *t*-haplotypes, mediated possibly through transmission ratio distortion.

The genus *Drosophila* maintains its special position at the forefront of empirical evolutionary biology, and the following four chapters help illustrate why. Aquadro and Begun explore genetic hitchhiking in the *Drosophila* genome. Combining information about recombination rate with the level of genetic variation from large-scale genomic surveys of restriction maps, they show that there is a significant negative correlation between the extent of DNA variation and the per-nucleotide recombination rate in different regions of the genome. They suggest that the hitchhiking effect leads to a reduction of average levels of standing polymorphism throughout a significant portion of the genome. Takano *et al.* analyze DNA sequences including the glycerol-3-phosphate dehydrogenase (*Gpdh*) gene from one *Drosophila simulans* and 11 *D. melanogaster* strains in relation to phenotypic variants, fast and slow alleles, as well as the latitudinal clines. They show that the evolutionary pattern of *Gpdh* is similar to that of *Adh*. An important issue in evolutionary biology centers on the genetic basis of reproductive isolation. Focussing on the development and genetics of spermatogenesis, Wu *et al.* review the feasibility of studying this phenomenon at the molecular level. As hoped, spermatogenesis may be responsible for erecting the first reproductive barrier between incipient species. A new technique, germ-plasm transplantation, has made it easy to produce hybrid flies which carry heterogeneous mitochondrial genomes derived from different subspecies of *D. melanogaster*. Matsuura finds that heteroplasmy thus produced exhibits selective transmission of mitochondrial genomes which is often dependent of temperature.

In contrast to sexually reproducing organisms, bacteria reproduce asexually and create recombinant genotypes only through mechanisms of gene transfer. Natural populations of bacteria harbor extensive genetic diversity in a limited number of genetically distinct clones. Whittam and Ake distinguish three types of past recombination events and summarize their relative importance in the generation of genotypic diversity in natural populations of *Escherichia coli*.

Some papers in this volume explicitly address questions concerning the genealogical history of genes and organisms, as the title of symposium indicates, and others less so. Nevertheless, all are fully cognizant of the importance of connection between changes at the DNA level and those at the phenotypic level. Such a common recognition is the basis of this volume, and it is our perception that we are witnessing the birth of a new and powerful mode of inquiry in evolutionary biology.

This publication was supported in part by grants from the Taniguchi Foundation and the Ministry of Education, Science and Culture, Japan.

August 1992   Naoyuki TAKAHATA
                                                       Andrew G. CLARK

Contents

ix

x

Evolutionary Genetics of Human Paleo-Populations

NAOYUKI TAKAHATA

National Institute of Genetics, Mishima 411, Japan

The pattern and time scale of the ancestral relationships among genes or alleles at a genetic locus are a reflection of molecular evolutionary mechanisms and evolutionary dynamics of their carrier population. If we know how alleles at a particular locus evolved and we reconstruct the ancestral relationships from the DNA sequence differences, we may be able to learn what might or might not have happened in the history of populations. Of particular interest is whether the founder effect (Mayr, 1963) has played an important role in speciation, or how geographically dispersed individuals were united to form a single species. My main concern here is with the evolutionary history of humans.

Like other orders in Mammalia, primates have a long history extending over about 70 million years (*e.g.*, Lewin, 1988). Primates were one of the groups of placental mammals that survived the mass-extinction event that occurred about 65 million years ago (mya). Living species of primates are often classified into the prosimians, the New and Old World monkeys, and the hominoids. The adaptive radiation of primates took place between 50 and 60 mya, followed by the 35-million-year-old-divergence of hominoids from the Old World monkeys. The latter date is based on the fossil of a generalized arboreal quadruped, *Aegyptopithecus* found in the Fayum Depression. It is believed that in the early Miocene, about 20 mya, apes had become extremely numerous and diverse. Of

1

all the Miocene apes, *Proconsul* is best known as a genealogical antecedent of modern apes and humans. But the resurgence of adaptive radiation and a deteriorating climate in the Miocene ended the Age of the Ape, leaving only the gibbon, orangutan, gorilla, chimpanzee, and humans. *Ramapithecus*, once considered as a hominid, lived 15 mya. There has been a dramatic shift in the view about the phylogenetic position of *Aegyptopithecus* and *Ramapithecus*. *Aegyptopithecus* is now considered not to be an ape, or a hominoid, or even a catarrhine primate, but to be linked to the common ancestral lineage leading to the New and Old World monkeys and hominoids. *Ramapithecus* is regarded as being linked to the common ancestral lineage for humans and African apes. Molecular data have supported the view that humans and apes have been distinct for only 5 million years (Sarich and Wilson, 1967), not 15 million years as previously thought (Pilbeam, 1984), and that the genetically closest relatives to humans are chimpanzees (Horai *et al.*, 1992 and references therein).

Even when we look at more recent history in the human lineage, there were several important evolutionary events. During the last 4 or 5 million years, there have been at least four or five different species on the direct line leading to *Homo sapiens sapiens*, although what happened before 4 million years is not known. An ape-like *Dryopithecus* lived about 4 mya from which *Australopithecus afarensis* was descended. *H. habilis* emerged from *A. afarensis* 2.5 mya and *H. erectus*, in turn, evolved from *H. habilis* 1.6 mya.

The first demonstrable migration of *H. erectus* took place from Africa 1.0–1.5 mya. An archaic form of *H. sapiens* evolved from *H. erectus* 0.3 mya and the modern human is believed to have emerged from an archaic form of *H. sapiens* about 0.1 mya, followed by racial differentiation. A debate on the origin of the modern human (*e.g.*, Howells, 1976; Smith and Spencer, 1987; Mellars and Stringer, 1989; Stringer, 1990) is centered on whether all living populations had a recent origin in the Late Pleistocene, some hundred thousand years ago, or whether they evolved in many different regions from local archaic populations of *H. erectus*. One hypothesis, called the candelabra, assumes no migration and parallel evolution of modern *H. sapiens* in several regional localities at the same time. An alternative, called the Noah's Ark, assumes the complete replacement of populations in the Old World by anatomically modern *H. sapiens* from Africa. In the former view racial differentiation is as old as the emergence of the modern human, while in the latter view it is relatively young, less than 0.2 million years. A modified version of the candelabra, called the multiregional hypothesis (Wolpoff *et al.*, 1987; Wolpoff, 1989), allows gene flow among different archaic populations of *H. erectus*. To distinguish these hypotheses, it is necessary to discern the role and extent of migrations which have occurred during the last 1.0–1.5 million years.

Thus, although man's place in nature has been a cardinal question ever since Darwin, we would also like to know more dynamical aspects in the evolution of human populations. Specifically, I ask whether a significant reduction in population size happened in a number of speciation events in the lineage leading to the modern human and what the geographic population structure looked like. To address these questions, we need to use various genetic systems and population genetics theory.

MOLECULAR DATA

There are two genetic systems in primates and, in particular, in human

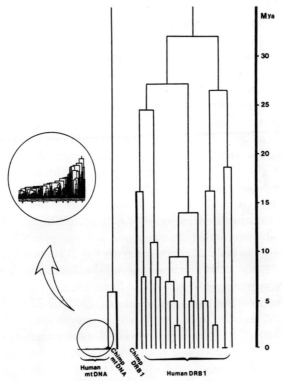

Fig. 1. Gene genealogy of mtDNA (Horai, 1991) and allelic genealogy at the *Mhc* class II *DRB1* in human populations (Satta *et al.*, 1991).

populations that have been intensively studied at the DNA level and which have evolved in quite different time scales: the mitochondrial genome (mtDNA) and the major histocompatibility complex (*Mhc*) loci. The ancestral relationships of human mtDNAs and those of *DRB1*, a class II locus of *Mhc*, are depicted in Fig. 1. The most recent common ancestor of mtDNAs from a worldwide survey existed relatively recently, no longer than 0.2 mya (Cann *et al.*, 1987; Kocher and Wilson, 1991; Horai, 1991; Vigilant *et al.*, 1991), whereas the *DRB1* alleles are much more ancient, probably arising around the time of divergence of hominoids from the Old World monkey, 35 mya (Satta *et al.*, 1991). Such a tremendous difference cannot be accounted for by the difference in the mode of inheritance (haploidy *vs.* diploidy and maternal *vs.* biparental) which changes the effective size (N) of a population only by a factor of four under usual circumstances. Rather, it is likely to be the result of whether or not natural selection has been involved in these genetic systems. If selection is absent and random drift is the sole evolutionary mechanism at the population level, the time scale of the ancestry of genes at a locus does not greatly exceed $4N$ generations (Fisher, 1930; Kingman, 1982). If the value of N in human populations is of the order of 10^4 and the average generation time is 20 years, $4N$ generations occur in less than one million years. Neutral genes in a population of effective size 10^4 are therefore capable of providing useful information about evolutionary events that occurred only within this relatively short time period.

The fact that the exhaustive sampling of mtDNAs from human populations could find no pair with a separation time longer than 0.2 million years (Horai, 1991; Vigilant *et al.*, 1991) signifies that recent human populations have not been strongly structured geographically. Were this the case, the effective size could have been much larger, and neutral gene genealogy is expected to be much longer than one million years (Takahata, 1991a). This near panmixia of human populations in the Late Pleistocene is also supported by F_{ST} analyses of electrophoretic variants at various enzyme loci. This is because F_{ST} equilibrates rapidly (Nei *et al.*, 1977; Crow and Aoki, 1984) and the estimated value, 0.31 for mtDNA (Stoneking *et al.*, 1990; Merriwether *et al.*, 1991) and 0.10 for nuclear enzyme loci (Nei and Roychoudhury, 1982), is fairly small.

By contrast, several *Mhc* loci are extremely polymorphic. More importantly, the age of *Mhc* alleles is incomparably old (for other similar genetic systems, see Ioerger *et al.* 1990; Clark and Kao, 1991; Clark, this volume; Hughes, this volume) and many existing alleles have been transmitted through many speciation events (*trans*-species mode of polymorphism; Klein, 1980, 1986). The time scale in the evolution of *Mhc* alleles is some tens of million years, at least

10 times greater than that of neutral genes. Therefore, polymorphic alleles at *Mhc* loci reflect the population dynamics of hominoids not only in the Pleistocene but also in the Miocene and Oligocene.

The main cause behind the generation of such a long ancestry of *Mhc* alleles is some form of natural selection (Hughes and Nei, 1988). Among a variety of forms of natural selection (Kimura, 1983), relevant to the evolution of *Mhc* alleles is the diversity-enhancing-type of selection like overdominance (Hughes and Nei, 1988; Takahata and Nei, 1990). However, there can be a class of frequency-dependent selection which is theoretically equivalent to overdominance selection (Takahata and Nei, 1990; Denniston and Crow, 1990). For this reason, Takahata *et al.* (1992) called such a class collectively overdominance-type selection (here, interchangeably, balancing selection). They examined the internal consistency of the model with the available DNA sequence data of *Mhc* alleles and found the model compatible with many aspects of the data (see also Satta, this volume). In the following, I shall present some theoretical results on the overdominance-type selection in relation to changes in population size and the geographic distribution of individuals in a species. However, since the genealogy of neutral genes is well investigated and useful as a reference, some results under neutrality are also presented.

GENEALOGY OF BALANCED ALLELES IN A FINITE POPULATION AT EQUILIBRIUM

We first compare the neutral gene genealogy (Kingman, 1982) with the ancestral relationships among alleles that have been subjected to balancing selection in a stationary population. If genes evolved according to a symmetric model of strong balancing selection, relatively weak mutation and drift, it is possible to construct a theory of allelic genealogy specifying the relationships among different allelic lines (Takahata, 1990). Figure 2 explains the topological relationships among such alleles. It turns out that the topology in an allelic genealogy can be generated according to the rule of random bifurcation of allelic lines because of the assumed symmetric balancing selection (Takahata *et al.*, 1992). The rate of allelic turnover or replacement of parental allele by descendant, however, depends on population size, selection intensity and mutation rate. Designate by s and u the selective advantage of heterozygotes over homozygotes and the mutation rate per target site of natural selection per generation, respectively. In the case of the Mhc molecule, such a target site of balancing selection is presumably the nonsynonymous sites of the peptide-binding region (PBR). We assume that a population of size N is panmictic in

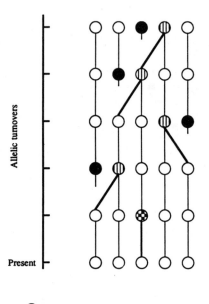

Allelic turnovers

Present

●   Extinct allelic lines

◍   Duplicated allelic lines

✪   Allelic lines replaced by their own descendants

Fig. 2. Diagram representing allelic turnovers under strong symmetric balancing selection and weak mutation in a finite population. The infinite allele model of Kimura and Crow (1964) is assumed.

which n allelic lines are maintained at equilibrium. We define coalescence time t_j as the time in the past at which j different alleles have been derived from $j-1$ distinct ancestral alleles for the first time. This time is a random variable and it was shown that the distribution of $t_j = t$ can be approximated by an exponential function (Takahata, 1990)

$$\frac{j(j-1)}{4Nf_s}\exp\left\{-\frac{j(j-1)t}{4Nf_s}\right\}, \qquad j=2,\ 3,\ \cdots,\ n \tag{1}$$

where

$$f_s \simeq \frac{\sqrt{S}}{2M}\left[\ln\left\{\frac{S}{16\pi M^2}\right\}\right]^{-3/2}, \tag{2}$$

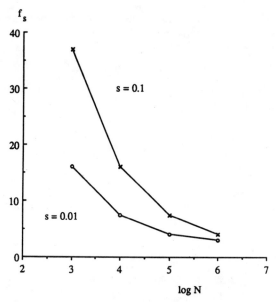

Fig. 3. Time scale of balanced allelic genealogy is $2Nf_s$. The scaling factor f_s is a function of N, mutation rate u ($=4.5 \times 10^{-6}$ per peptide-binding region per generation here) and selection coefficient s.

$M = Nu$ and $S = 2Ns$. In terms of f_s, the mean number of alleles maintained is expressed by

$$n^3 = 4\sqrt{2}MSf_s \tag{3}$$

(Kimura and Crow, 1964). Because t_j is exponentially distributed, the allelic genealogy becomes mathematically the same as the genealogy of neutral genes, except for the time scale or allelic turnover rate. An appropriate scaling factor in allelic genealogy is $2Nf_s$, which can be much larger than $2N$ for small M and large S (Fig. 3).

Define T_n as $\sum_{j=2}^{n} t_j$. The T_n is the time in the past at which the most recent common ancestor of all n alleles existed for the first time. The mean and variance of T_n follow immediately from Eq. 1 as

$$E\{T_n\} = 4Nf_s\left(1 - \frac{1}{n}\right) \quad \text{and} \quad V\{T_n\} = (4Nf_s)^2 \sum_{j=2}^{n} \frac{1}{\{j(j-1)\}^2}, \tag{3}$$

respectively.

It follows from the distribution of t_j in Eq. (1) that the probability $g_{nk}(t)$

that there were k distinct ancestral allelic lines t generations ago for a sample of n alleles can be approximated by

$$g_{nk}(t) = \sum_{j=k}^{n} \frac{(-1)^{j-k}(2j-1)k_{(j-1)}n_{[j]}}{k!(j-k)!n_{(j)}} \exp\left\{-\frac{j(j-1)t}{4Nf_s}\right\}$$ (4a)

for $2 \le k \le n$, and

$$g_{n1}(t) = \sum_{j=1}^{n} \frac{(-1)^{j-1}(2j-1)n_{[j]}}{n_{(j)}} \exp\left\{-\frac{j(j-1)t}{4Nf_s}\right\}$$ (4b)

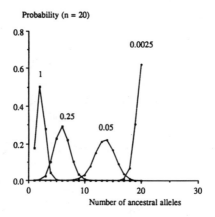

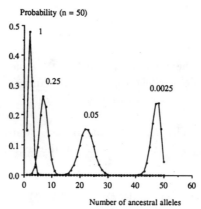

Fig. 4. The probability $g_{nk}(t)$ in Eq. (4) (ordinate) and the number (k) of distinct ancestral genes in a random sample of size n (in abscissa). Time, $t/(2Nf_s)$, is presented beside each broken line.

for $k=1$, where $a_{(j)}=a(a+1) \ldots (a+j-1)$ and $a_{[j]}=a(a-1) \cdots (a-j+1)$ for $j \geq 1$, and $a_{(0)}=a_{[0]}=1$. Equation (4) with $f_s=1$ is Tavaré's (1984) expression of probability $g_{nk}(t)$ for a sample of n neutral genes. Figure 4 depicts this probability as a function of n and $t/(2Nf_s)$. Clearly, coalescence of balanced alleles occurs in the time scale of $2Nf_s$ generations (Takahata, 1990).

BOTTLENECK AND FOUNDER PRINCIPLE

Since a large number of alleles can be maintained only when N is sufficiently large, it is interesting to study what happens when N becomes small. We model this situation by an abrupt bottleneck. The ancestral population is assumed to be at equilibrium in which there are n different alleles with frequency p_j $(j=1, 2, \cdots, n)$. Suppose that a stationary population abruptly decreased in size t generations ago. For simplicity, as in a panmictic population, we use the same symbol N to represent the reduced population size. We assume that the founder population consists of $2N$ genes randomly sampled from the ancestral, stationary population. Otherwise, we may assume that the founder population is an exact miniature of the ancestral population, receiving all existing alleles with frequencies p_j. The latter assumption becomes equivalent to the former after random sampling of gametes in the first generation. Founder alleles in a sample are called old alleles if they do not mutate during t generations, or nonmutant lines-of-descent by Griffiths (1980).

We ask how many old alleles have persisted for t generations. There are two situations of particular interest: (a) the Ns is so small that balanced alleles behave as neutral owing to reduced N and (b) the Ns is still so large that selection predominates and allele frequencies are in balance.

In the first case of neutrality, Tavaré (1984) found useful formulas for the probability that there were r distinct alleles t generations ago and the probability distribution of time T_r at which for the first time there are exactly r distinct alleles in the population $(r=1, 2, \cdots, n-1)$. Although these formulas were derived under the assumption that a population is at equilibrium, they can directly be applied to the present model of bottlenecks (see also Watterson, 1984). For instance, the mean of T_r in the absence of mutation is

$$E\{T_r\}=4N\sum_{k=1}^{r}(-1)^{r-k+1}\binom{n-1-k}{r-k}\sum_k\{1-p_{i1}-\cdots-p_{ik}\}\ln(1-p_{i1}-\cdots-p_{ik}) \quad (6)$$

where $\sum_k$ is the sum over all choices $1 \leq i_1 \leq i_2 \leq \cdots \leq i_k \leq n$ and we recall that p_{ik} stands for the frequency of the i_k allele in the founder population.

TABLE I

Mean and Standard Deviation ($\pm$) of T_r in Units of $2N$ Generations ($n=10$) in a Population under Bottleneck

	$r=9$	8	7	6	5	4	3	2	1
Eq. 7	0.081	0.120	0.163	0.216	0.286	0.389	0.559	0.894	1.90
Sim.	0.075	0.111	0.154	0.209	0.281	0.386	0.544	0.874	1.87
$\pm$	0.030	0.039	0.053	0.071	0.102	0.142	0.209	0.365	1.04

Sim.: simulation results of 500 repeats. $N=200$.

When the ancestral population is subjected to symmetric balancing selection, mutation, and random sampling drift, the allele frequencies are more or less equal to each other. Thus the allele frequencies in the founder population all are assumed to be $1/n$. This is a conservative assumption to evaluate the effect of bottleneck because it provides the most polymorphic situation. Equation (6) then becomes

$$E\{T_r\}=4N\sum_{k=1}^{r}(-1)^{r-k+1}\binom{n-1-k}{r-k}\binom{n}{k}\left(1-\frac{k}{n}\right)\ln\left(1-\frac{k}{n}\right)$$

$$=4N\sum_{k=1}^{r}(-1)^{r-k+1}\frac{(n-1)!}{k!(r-k)!(n-r-1)!}\ln\left(1-\frac{k}{n}\right)$$

(7)

(Table I). The mean time of $T_{r=1}$ is especially simple, and is $E\{T_{r=1}\}=4N(n-1)\ln\{n/(n-1)\}$ (Litter, 1975). For large n, it becomes approximately $4N$, corresponding to the well-known result for the mean fixation time of a neutral allele (Kimura and Ohta, 1969).

Klein *et al.* (1990) performed a computer simulation to observe how quickly balanced alleles are lost in a small population (ignoring the effect of new mutations). To study the time this process requires, we use a diffusion approximation for the change of gene frequency x of a particular allele. Suppose that loss of old alleles occurs one after another and that j old alleles ($1 \leq j \leq n$) are retained at a certain time in the population under bottleneck. The mean change of x per $2N$ generations and the variance are given respectively by

$$a(x)=-2Mx-Sx(x-F_j) \quad \text{and} \quad b(x)=x(1-x)$$

(8)

where s is assumed to be small, $M=Nu$, $S=2Ns$ and F_j is the homozygosity due to these j balanced alleles. Because of the symmetric balancing selection, we approximate F_j by $(1/j)(1-x)^2+x^2$. ($F_j \approx 1/j$ leads to essentially the same result.) Then the mean time until a particular allele becomes lost may be given by

$$E\{T_j(x)\}=4N\left[\int_0^x \frac{dy}{b(y)\psi(y)}\int_0^y \psi(z)dz+\int_x^1 \frac{dy}{b(y)\psi(y)}\int_0^x \psi(z)dz\right] \tag{9}$$

where $\psi(x)=(1-x)^{-4M}\exp\{S^*(x-m_j)^2)\}$, $S^*=jS/(j-1)$ and $m_j=1/j$. An approximate solution of Eq. 9 with $x=m_j$ is

$$E\{T_j(m_j)\}\approx \frac{2N\sqrt{\pi}\{\exp(S^*m_j^2)-\exp(-S^*m_j^2)\}}{2S^*\sqrt{S^*}\,m_j^2(1-m_j)^{1-4M}}\{\mathrm{erf}\sqrt{S^*}\,m_j+\mathrm{erf}\sqrt{S^*}(1-m_j)\}.$$

$$\tag{10}$$

When there exist j segregating alleles, the mean time until one of the j alleles

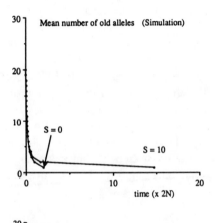

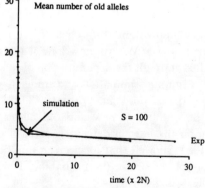

Fig. 5. The number of old alleles retained in a population (ordinate) as a function of time since bottleneck started (abscissa). See also Klein *et al.* (1990) for simulation study of this problem.

is lost, $E\{U_j\}$, should be given by dividing $E\{T_j(m_j)\}$ by j. That is

$$E\{U_j\} \approx \frac{1}{j} E\{T_j(m_j)\},$$

so that the mean time of T_r at which there are $r(r = 1, 2, \cdots, n-1)$ old alleles for the first time is

$$E\{T_r\} \approx \sum_{j=r+1}^{n} E\{U_j\}. \tag{11}$$

If $1 \ll S \ll j(j+1)$, the value of $E\{T_r\}$ is small. Therefore, even when balancing selection operates, extensive loss of founder alleles may possibly occur within a short time period (Fig. 5; Klein *et al.*, 1990); the situation is rather similar to that for neutral alleles. However, as the value of j approaches the number of alleles that can be maintained in a population of reduced size, there is a large difference in $E\{U_j\}$, depending on whether alleles are neutral or balanced. Since balancing selection is assumed to be effective, $E\{T_r\}$ in Eq. (11) can be enormous when a few alleles are segregating. If $E\{T_r\} \gg 2N$ for some r, it is likely that r such old alleles may escape from extinction during the bottleneck phase. For instance, the value of r for $E\{T_r\} \gg 2N$ may be 4 when $n = 20$, $S = 100$ and $M = 0.001$, because $E\{T_{r=4}\} \approx 55N$ from Eq. (11) and $40 \pm 28N$ by simulation. Nevertheless, a large number of alleles cannot survive through a bottleneck even if they are subjected to fairly strong balancing selection. Mutation has a rather small effect, and tends to reduce the persistence time of founder alleles.

POPULATION EXPANSION

Suppose that a population went through a severe bottleneck and afterward abruptly expanded in size. We ask how long it takes for a new genetic equilibrium to be reached through the accumulation of newly arisen mutations (new alleles). Based on computer simulation and under neutrality, Maruyama and Fuerst (1984) studied this problem. Watterson (1984) provided analytic formulas concerning the number of new and old alleles after a bottleneck, basing their derivation on the lines-of-descent process of Griffiths (1980). For example, Watterson (1984) showed that for a given number (k) of nonmutant lines-of-descent in a sample of size n, the probability generating function for the number of new alleles is given by

$$\frac{(4Mz + k)_{(n-k)}}{(4M + k)_{(n-k)}} = \prod_{j=k}^{n-1} \frac{4Mz + j}{4M + j} \tag{12}$$

where $M = Nu$ (in this section, N stands for the size of the expanded population) and z is a dummy variable in the probability generating function. The mean number of new alleles is

$$\prod_{j=k}^{n-1} \frac{4M}{4M+j}.$$

The above formula is identical to that for a stationary population (Ewens, 1972), if $k=0$ with probability 1. As the probability of taking large values of k becomes high for a short elapsed time since population expansion, the mean number of new alleles becomes small for obvious reasons. It should be noted that the number of nonmutant lines-of-descent k is determined by an equation analogous to Eq. (4) (Griffiths, 1980; Tavaré, 1984). Tables I and II in Watterson (1984) show that, unless $M \ll 1$, a new equilibrium state is attainable, in most cases, nearly within $2N$ generations after population expansion.

We assume that the expanded population was initially monomorphic. Suppose that the jth new polymorphic allele is introduced into the population by mutation so that there are $j+1$ balanced alleles, one founder and j new alleles. The number of j is assumed to be no greater than the number of alleles that can be maintained in the expanded population at equilibrium (Eq. (3)). It should be noted that strongly balanced alleles are unlikely to become fixed. Therefore, we instead compute the partial fixation probability that the jth allele with initial frequency $x = 1/(2N)$ reaches frequency $m_j = 1/(1+j)$ before extinction. For this purpose, we slightly modify Kimura's (1962) formula for the fixation probability as

$$u_j(x) = \frac{\int_0^x f_j(y)dy}{\int_0^{m_j} f_j(y)dy}. \tag{13}$$

In the above, $f_j(y)$ is given by $\exp\{S^*(y-m_j)^2\}$ with $S^* = S(j+1)/j$ under the present model of balancing selection. If S^* is large, $u_j(1/(2N))$ can be approximated as

$$u_j\left(\frac{1}{2N}\right) = \frac{1 - \exp\{-\dfrac{2S^* m_j}{2N}\}}{1 - \exp\{-2S^* m_j^2\}}, \tag{14}$$

so that the rate of incorporation of the jth mutation into the non-equilibrium population is given by $2Mu_j(1/(2N))$ ($M = Nu$). The mean incorporation time of r new alleles becomes

TABLE IIa

The Mean and Standard Deviation ($\pm$) of the Number of New Alleles at $t/(2N)$ Units of Time after Starting from Zero Variability ($Nu=0.1$)

$t/(2N)$	0.1	0.2	0.3	0.4	0.5	0.6	0.7	0.8	0.9
$S=0$	1.45	1.74	1.92	2.01	2.10	2.23	2.14	2.34	2.38
$\pm$	1.18	1.32	1.38	1.34	1.41	1.52	1.45	1.51	1.51
$S=50$	2.52	3.65	4.26	4.94	5.24	5.60	5.85	6.12	6.34
$\pm$	1.53	1.65	1.71	1.78	1.73	1.68	1.72	1.80	1.78
$S=500$	6.66	9.37	11.2	12.6	13.7	14.1	14.6	14.8	15.0
$\pm$	1.91	2.13	2.16	2.27	2.28	2.10	2.07	1.98	2.01

$N=500$ and the number of replicates is 500 in simulation.

TABLE IIb

The Mean Time (in Units of $2N$ Generations) until r New Alleles Are Incorporated into an Expanded Population (Eq. 15)

	$r=1$	2	3	4	5	6	7	8	9
$S=50$	0.051	0.152	0.303	0.502	0.744	—	—	—	—
$S=500$	0.006	0.018	0.034	0.054	0.080	0.110	0.146	0.186	0.231

$$E\{T_r\} \approx \frac{1}{2M} \sum_{j=1}^{r} \frac{1-\exp\{-\dfrac{2S}{j(j+1)}\}}{1-\exp\{-\dfrac{S}{jN}\}}. \tag{15}$$

This approximation is in good agreement with simulation results and shows that for large S and M and for small r, $E\{T_r\}$ is of the order of $1/(4MS)$ in units of $2N$ generations. For instance, $E\{T_{r=1}\} \approx 0.05N$ for $M=0.1$ and $S=100$ while $E\{T_{r=1}\} \approx 0.005N$ for $M=1$ and $S=100$. This indicates that it does not take long for a new equilibrium to be attained after population expansion (Fig. 6, Table II).

Thus, even if a bottleneck occurred recently and a population became monomorphic at balanced loci, there can be a large number of balanced alleles shortly after population expansion. Most new alleles, however, cannot be old and must be closely related at the DNA sequence level. It is therefore reasonable to conclude that if a population experienced a severe constriction and then expansion in size, the genealogy of balanced alleles must exhibit a cluster of coalescences within a short time period, something similar to the shape of a star phylogeny.

The allelic genealogy at the *Mhc* inferred from DNA sequence differences (Fig. 1) does not show such a conspicuous shape, except for an indication of a rapid coalescence around the early Miocene divergence of apes and the

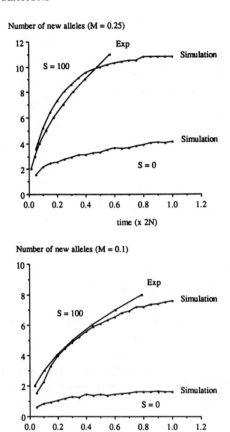

Fig. 6.   The number of newly arisen alleles since a population abruptly expanded.

divergence of African apes. This suggests that, for most of the time, the human lineage has not undergone any severe constriction and expansion in its population number. In particular, there is no such indication at the time the modern human emerged, contrary to the Noah's Ark hypothesis. More generally, since there have been many other speciation events in the human lineage during the last some tens of million years, the *Mhc* allelic genealogy is clearly incompatible with the founder principle as being important in human evolution (Takahata, 1990).

POPULATION STRUCTURE

We now address two problems concerning the effect of geographically structured populations. One is related to the multiregional hypothesis for the origin of modern humans. If demes or subpopulations are strongly isolated, even an advantageous mutation takes a long time to spread over the whole population. And if such a mutation played an important role in the evolution of modern humans, to what extent was gene flow necessary for that mutation to spread within one million years or so? The other problem is related to the extraordinary polymorphism at *Mhc* (Klein, 1986). The question is then whether or not population structure had a significant effect on the *Mhc* polymorphism.

1. Spread of Advantageous Mutations

We assume that the population structure is modeled by a finite island model of L demes, each with size N (not to be confused with the previous N), and that initially there is one deme fixed by a mutant gene. There are two possibilities; first, a deme once fixed by a mutant gene may be swamped out again by nonmutant immigrants from other demes and, second, the mutant gene may propagate to other demes and become fixed therein. The number of demes fixed by an advantageous mutation is a random variable and may be approximated by a birth and death process, provided that either "fixation" in one of the nonmutant demes or "extinction" in one of the mutant demes can happen rapidly (Slatkin, 1981; Takahata, 1991b).

Takahata (1991b) obtained the fixation probability in the entire population when there are initially i demes fixed by an advantageous mutation. For genic selection and small migration rate m, the fixation probability (denoted by V_i) becomes

$$V_i = \frac{1 - e^{-2Si}}{1 - e^{-2SL}}, \quad \text{for} \quad 1 \le i \le L - 1 \tag{16}$$

in which $S = 2Ns$ and s is the selection intensity of an advantageous allele. We may compare Eq. (16) with the formula of Kimura (1962) for the fixation probability in a panmictic population; demes are equivalent to individuals in a panmictic population.

We can also compute the mean time at which there are j demes fixed by an advantageous allele, provided that there are initially i such demes ($1 \le i \le j \le L$) (*e.g.*, pp. 69-74 in Ewens, 1979). From this mean time and again for genic

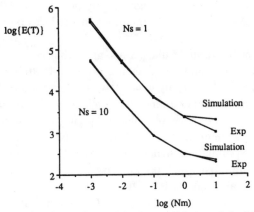

Fig. 7. The mean spreading time of a semidominant allele in a finite island model of 100 demes. The time and gene flow (Nm) are presented in the common logarithmic scale.

selection, we have the mean fixation time in the entire population as (Takahata, 1991b)

$$E\{T_i\} = \frac{(L-1)(1-e^{-2S})}{2Sm(1-e^{-2SL})} \times$$

$$\left\{(e^{-2Si} - e^{-2SL})\sum_{j=1}^{i}\frac{e^{2Sj}-1}{j(L-j)} + \sum_{j=i+1}^{L-1}\frac{(e^{-2Sj}-e^{-2SL})(e^{2Sj}-1)}{j(L-j)}\right\} \tag{17a}$$

for $i \geq 2$, and

$$E\{T_1\} = \frac{(L-1)(1-e^{-2S})}{2Sm(1-e^{-2SL})}\sum_{j=2}^{L-1}\frac{(e^{-2Sj}-e^{-2SL})(e^{-2Sj}-1)}{j(L-j)}, \tag{17b}$$

When S is large, $E\{T_i\} \ll (L-1)/(2SmL)\sum_{j=i}^{L-1}(1/j)$ and $E\{T_1\} \approx 1/(2Sm)$ $\{0.5771+\ln(L-1)\}$ for large L. These formulas are based on various approximations but are in good agreement with simulation results (Fig. 7).

It is clear that it takes a long time even for advantageous mutations to spread over the whole population if $Nm \ll 1$. The dependency on L is rather weak. We may imagine that new mutations, essential to modern human evolution, appeared within the last one million years (amounting to 0.5 million generations) and spread through the entire human population. If this is the case, we can use it to set the lower bound of the value of Nm. If Nm is as small as 0.001, a mutation even when strongly favored takes nearly 0.1 million generations or 2 million years to spread (Fig. 7). This does not meet the above requirement, so that Nm must be significantly larger than 0.001. A reasonable lower bound of Nm for human populations in the last one million years may be 0.01.

On the other hand, molecular data suggest that racial differentiation took place in the last 0.1 million years (Nei and Livshits, 1990). For this time period, the estimated level ($Nm \approx 1$) of gene flow from both mtDNA and nuclear enzyme loci is much higher than 0.01. Such an estimate seems reasonable, because the mtDNA sequence and allele frequency data both are by and large consistent with the neutral theory (Kimura, 1968, 1983; Nei, 1987), and a strongly isolated population structure makes it difficult for the ancestry of neutral genes to date back only 0.2 million years (Takahata, 1991b); the value of Nm cannot be as small as 0.01. If $Nm \approx 1$, however, racial characteristics must be accounted for by fairly strong local selection that could overcome the moderate level of gene flow.

2. Mhc Polymorphism

We have argued that human population structure prior to the Pleistocene may not be able to be inferred from the genealogy of neutral genes. It is interesting to ask whether *trans*-species polymorphism of *Mhc* is informative in this respect. Balancing selection maintains segregating alleles for a long time so that they reflect the early history of human populations.

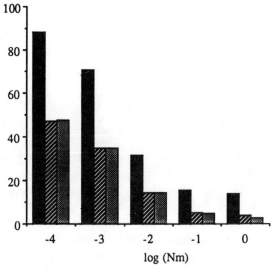

Fig. 8. The number of alleles in a finite island model of 50 demes at equilibrium. Solid bars for $Ns=5$, hatched bars for $Ns=0.2$, and shaded bars for $Ns=0$.

If a population is divided into many small demes, balancing selection becomes less efficient. Nevertheless, such inefficiency may be offset by increased neutral variation independently accumulating in isolated demes. There seems to be no satisfactory theory about the effect of population subdivision on balanced alleles and apparently there is a difficulty in its rigorous mathematical treatment. I therefore conducted a computer simulation to observe the joint effects of population subdivision and balancing selection on the number of alleles.

My main conclusion is that, if selection is effective within demes ($Ns > 1$) and the number of demes is large ($L > 10$), the interplay between balancing selection and population structure can produce an enormous number of alleles. This is because each deme can be polymorphic by selection and different demes tend to maintain different alleles owing to restricted gene flow (Fig. 8). This may help to explain why there are so many alleles at some *Mhc* loci. The condition required for gene flow is rather strict, however ($Nm \ll 0.1$). Population subdivision is only able to increase substantially the number of balanced alleles, when gene flow is limited to that extent.

This may appear incompatible with the mtDNA sequence and allele frequency data, but these data do not necessarily rule out the possibility that human populations prior to the Late Pleistocene were highly structured geographically. In fact, *Mhc* polymorphisms suggest that the long-term effective population size might have been of the order of 10^5 (Takahata *et al.*, 1992). Such a large value is incompatible with the estimate of 10^4 based on (presumably) neutral loci (Nei, 1987). It is, however, necessary to remember that balanced alleles inscribe documents of tens of million years in hominoid evolution while neutral alleles are those that are likely to be less than one million years old. One possibility is therefore that the dispersal rate was low before the Late Pleistocene but became high later on. Such a change can *decrease* the effective size from 10^5 to 10^4 (Nei and Takahata, unpublished). If this transition really occurred after the first migration of *H. erectus* from Africa, we can expect that the genealogy of neutral genes is an evolutionary document only from that time on, whereas the genealogy of *Mhc* alleles is still largely a reflection of the earlier population history in the human lineage.

SUMMARY

Genetic variation revealed at the DNA level provides useful information about not only the evolutionary mechanisms involved but also the history of organisms. As to the former, we have learned, at least in principle, how and what kind of genetic variation is generated at the molecular level and how it is shaped by the mechanisms acting at the population level. An important task is to clarify outcomes of simultaneously interacting molecular and population mechanisms and to assess their relative importance in actual evolutionary processes. Another exciting finding emerging in current evolutionary biology is the possibility that genetic variation, together with appropriate theories, may shed new light on the population history of organisms and population dynamics including speciation processes. In this pursuit, it is important to delimit the time scale in which genetic variants have persisted and therefore are useful as a document of population histories. Equally important, to interpret these data, is development of appropriate theories. Focussing on these two points, I present some theoretical results and apply them to the major histocompatibility complex loci in humans. I argue that the human lineage has never experienced severe bottlenecks and that the extent of panmixia of the population might have been low before humans migrated out of Africa but later became high. The implication is related to the origin of modern humans, subsequent racial differentiation and the founder principle, a popular mechanism often invoked in speciation.

REFERENCES

Cann, R.L., M. Stoneking, and A.C. Wilson, 1987. *Nature* **325**: 31-36.

Clark, A.G. and T.-H. Kao, 1991. *Proc. Natl. Acad. Sci. U.S.A.* **88**: 9823-9827.

Crow, J.F. and K. Aoki, 1984. *Proc. Natl. Acad. Sci. U.S.A.* **81**: 6073-6077.

Denniston, C. and J.F. Crow, 1990. *Genetics* **125**: 201-205.

Ewens, W.J., 1972. *Theor. Popul. Biol.* **3**: 87-112.

Ewens, W.J., 1979. *Mathematical Population Genetics.* Springer-Verlag, Berlin.

Fisher, R.A., 1930. *The Genetical Theory of Natural Selection.* Dover Publications, New York.

Griffiths, R.C., 1980. *Theor. Popul. Biol.* **17**: 37-50.

Horai, S., 1991. In *New Aspects of the Genetics of Molecular Evolution*, edited by M. Kimura and N. Takahata, pp. 135-152. Japan Sci. Soc. Press, Tokyo/Springer-Verlag, Berlin.

Horai, S., Y. Satta, K. Hayasaka, R. Kondo, T. Inoue, T. Ishida, S. Hayashi, and N. Takahata, 1992. *J. Mol. Evol.* **35**: 32-43.

Howells, W.W., 1976. *J. Hum. Evol.* **5**: 477-495.

Hughes, A.L. and M. Nei, 1988. *Nature* **335**: 167-170.

Ioerger, T.R., A.G. Clark, and T.-H. Kao, 1990. *Proc. Natl. Acad. Sci. U.S.A.* **87**: 9732-9735.

Kimura, M., 1962. *Genetics* **47**: 713-719.

Kimura, M., 1968. *Nature* **217**: 624-626.

Kimura, M., 1983. *The Neutral Theory of Molecular Evolution.* Cambridge University Press, Cambridge.

Kimura, M. and J.F. Crow, 1964. *Genetics* **49**: 725-738.

Kimura, M. and T. Ohta, 1969. *Genetics* **61**: 763-771.

Kingman, J.F.C., 1982. *J. Appl. Probab.* **19A**: 27-43.

Klein, J., 1980. In *Immunology 80*, edited by M. Fougereau and J. Dausset, pp. 239-253. Academic Press, London.

Klein, J., 1986. *Natural History of the Major Histocompatibility Complex*. John Wiley & Sons, New York.

Klein, J., J. Gutknecht, and N. Fischer, 1990. *Trends Genet.* **6:** 7-11.

Kocher, T.D. and A.C. Wilson, 1991. In *Evolution of Life: Fossils, Molecules and Culture*, edited by S. Osawa and T. Honjo, pp. 391-413. Springer-Verlag, Berlin.

Lewin, R., 1988. *In the Age of Mankind.* Smithsonian Book, Washington D.C.

Litter, R.A., 1975. *Math. Biosci.* **25**: 151-163.

Mayr, E., 1963. *Animal Species and Evolution.* Harvard Univ. Press, Cambridge, Massachusetts.

Maruyama, T. and P.A. Fuerst, 1984. *Genetics* **108**: 745-763.

Mellars, P. and C. Stringer, 1989. *The Human Revolution: Behavioral and Biological Perspectives on the Origin of Modern Humans*. Princeton University Press, Princeton, New Jersey.

Merriwether, D.A., A.G. Clark, S.W. Ballinger, T.G. Schurr, H. Soodyall, T. Jenkins, S.T. Sherry, and D.C. Wallace 1991. *J. Mol. Evol.* **33**: 543-555.

Nei, M., 1987. *Molecular Evolutionary Genetics.* Columbia University Press, New York.

Nei, M., A. Chakravarti, and Y. Tateno, 1977. *Theor. Popul. Biol.* **11**: 291-306.

Nei, M. and G. Livshits, 1990. In *Population Biology of Genes and Molecules*, edited by N. Takahata and J.F. Crow, pp. 251-265. Baifukan, Tokyo.

Nei, M. and A.K. Roychoudhury, 1982. *Evol. Biol.* **14**: 1-59.

Pilbeam, D., 1984. *Sci. Am.* **250**: 84-96.

Sarich, V.M. and A.C. Wilson, 1967. *Science* **158**: 1200-1203.

Satta, Y., N. Takahata, C. Schöach, J. Gutknecht, and J. Klein, 1991. In *Molecular Evolution of the Major Histocompatibility Complex*, edited by J. Klein and D. Klein, pp. 51-62. Springer-Verlag, Heidelberg.

Slatkin, M., 1981. *Evolution* **35**: 477-488.

Smith, F.H. and F. Spencer, 1987. *The Origins of Modern Humans.* Alan R. Liss, Inc., New York.

Stoneking, M., L.B. Jorde, K. Bhatia, and A.C. Wilson, 1990. *Genetics* **124**: 717-733.

Stringer, C., 1990. *Sci. Am.* **263**: 98-104.

Takahata, N., 1990. *Proc. Natl. Acad. Sci. U.S.A.* **87**: 2419-2423.

Takahata, N., 1991a. In *Molecular Evolution of the Major Histocompatibility Complex*, edited by J. Klein and D. Klein, pp. 29-49. Springer-Verlag, Heidelberg.

Takahata, N., 1991b. *Genetics* **129**: 585-595.

Takahata, N. and M. Nei, 1990. *Genetics* **124**: 967-978.

Takahata, N., Y. Satta, and J. Klein, 1992. *Genetics* **130**: 925-938.

Tavaré, S., 1984. *Theor. Popul. Biol.* **26**: 119-164.

Vigilant, L., M. Stoneking, H. Harpending, K. Hawkes, and A.C. Wilson, 1991. *Science* **253**: 1503-1507.

Watterson, G.A., 1984. *Theor. Popul. Biol.* **26**: 77-92.

Wolpoff, M.H., W.X. Zhi, and A.G. Thorne, 1987. In *The Origin of Modern Humans*, edited by F.H. Smith and F. Spencer, pp. 411-483. Alan R. Liss, Inc., New York.

Wolpoff, M.H., 1989. In *The Human Revolution*, edited by P. Mellars and C. Stringer, pp. 62-108. Princeton University Press, Princeton, New Jersey.

The How and Why of Generating Gene Genealogies

RICHARD R. HUDSON

Department of Ecology and Evolutionary Biology, University of California, Irvine, CA 92717, U.S.A.

Consideration of gene genealogies and the use of the coalescent process in analyzing models without recombination have become commonplace. However, for models with recombination between two or more loci, the methods are not so well known. In this paper I will review briefly the coalescent process for one locus and describe the use of the coalescent process in computer simulations. A distinct method of generating the distribution of test statistics for the neutral model will be described. I will then describe the coalescent process for a two-locus neutral model, and describe how it can be used in computer simulations to estimate the joint distribution of allele frequencies and linkage disequilibrium between two loci. The use of this distribution to interpret data and to estimate the recombination parameter of a neutral model will also be described.

ONE LOCUS MODEL

Figure 1 shows an example of a gene genealogy of a sample of five alleles from a population. Under a Wright-Fisher neutral model with a constant population size of N diploids, the times between the nodes of the genealogy are exponentially distributed with means that depend on the number of lineages being considered and the population size. The topology of the tree under this

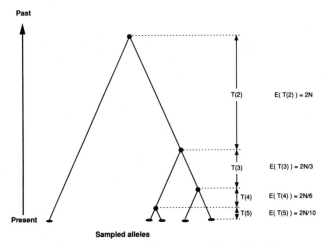

Fig. 1.   An example of a gene genealogy of a sample of five alleles. The expected durations of the different time intervals under a Wright-Fisher neutral model are shown on the right, measured in units of generations.

model can be generated by randomly choosing lineages to combine at each node. Genealogies under this model are very simple and fast to produce on a computer.

If we assume that in each generation in each lineage there is a small probability, u, that a mutation occurs, then the expected number of mutations on a given gene genealogy is uT, where T is the sum of the lengths of the branches of that gene genealogy. If the occurrence of mutations at different times and in different lineages are independent of each other, then the number of mutations on a genealogy of given size is Poisson distributed. Given a genealogy and the number of mutations that have occurred on the genealogy, the distribution of the mutations on the genealogy is simple to describe. Specifically, given that a mutation has occurred somewhere on the genealogy the probability that it has occurred on a particular branch is proportional to the length of that branch. Under an infinite-site model, in which each mutation occurs at a previously unmutated site, the haplotypes produced are determined solely by the number of mutations and their location on the gene genealogy. At equilibrium under this model, all the sampling properties depend on n, the sample size, and $\theta = 4Nu$.

The preceding two paragraphs are an outline of a computer algorithm for generating samples under a Wright-Fisher neutral model. The basic steps are: (1) generate the times of the nodes of the genealogy, (2) generate the topology of the genealogy, (3) generate S, the number of mutations that occur on the genealogy (conditional on the size of the genealogy), (4) sprinkle S mutations on the genealogy, and (5) produce the resulting gametes. An implementation of this algorithm can be seen in Hudson (1990). This type of program is simple and fast and can be used to study the statistical properties of sampled sequences. In the next few paragraphs I will illustrate the utility of such computer programs, by considering the problem of determining the distribution of a test statistic, D, proposed by Tajima (1989) for testing the neutral model.

To make this illustration more concrete, I consider a specific hypothetical data set. I suppose that a sample of 10 sequences of a genomic region are obtained, that 10 segregating sites are observed in this sample, and that the mean number of differences between pairs of sequences is 4.93. With these numbers, D can be calculated and it is 1.75. I then ask: should the equilibrium neutral model be rejected for this gene region? Supposing that I had no prior alternative hypothesis in mind, it would be reasonable to reject the neutral model if the observed value of D was outside of a 95% confidence region for D. That is, I would reject if the probability of $D \geq 1.75$ is less than 0.025, or the probability of $D \leq 1.75$ is less than 0.025. The mean value of D under the neutral model is approximately zero so it is only the probability $D \geq 1.75$ that is relevant for our example.

There are several different distributions of D that one might estimate by computer simulation and use for the interpretation of actual data. One distribution of D that could be used to address this question would be obtained by assuming a value of θ, and then generating many samples as described above and tabulating the resulting D values. I will refer to this as the "distribution of D given θ." Tajima (1989) displayed estimates of this distribution of D given θ obtained from computer simulations. To carry out our test of neutrality for the hypothetical data set using the distribution of D given θ, we need to estimate the probability that $D \geq 1.75$ under this distribution. This I have done using simulations as described above and using the following obvious estimator:

$$<\mathrm{Prob}(D \geq 1.75)> = \sum_{i=1}^{m} \chi(D_i \geq 1.75)/m,$$

where D_i is the value of D for the ith sample generated in the simulations, and $\chi(D_i \geq 1.75)$ is one if D_i is greater than or equal to 1.75 and zero otherwise, and m is the number of replicate samples generated. These simulations show that this probability has some dependence on θ. For our example, I find that the probability of D being greater than or equal to 1.75 is 0.031, 0.025, and 0.020

for θ equal to 1.77, 3.53, and 7.07, respectively. (These estimated probabilities as well as the other estimates in this section are based on 100,000 replicates.) These values of θ were chosen to be compatible with the number of segregating sites observed in our hypothetical sample. When $\theta = 3.53$, the average number of segregating sites in a sample of size 10 is 10. But a value of θ half this large (1.77) or twice this large (7.07) is not incompatible with the observation of 10 segregating sites. (Given that the average estimate of θ per base pair is about 0.005 for *Drosophila melanogaster*, a region about 700 base pairs long in this species corresponds to a value of θ equal to 3.5.) Thus, our hypothetical data suggest rejection of the neutral model for two of the values of θ and no rejection for the other value of θ. Since θ is typically unknown, this presents us with some difficulty in interpreting our data.

In an attempt to resolve this problem, Dr. Charles Aquadro and I have investigated two other distributions of D. We noted that the number, S, of segregating sites varies from one sample to the next in the above simulations. The distribution of S depends on θ. We thought it possible that the dependence of the distribution of D on θ might be through its dependence on S, and that therefore the distribution of D conditional on the value of S, might be independent or nearly independent of θ.

This distribution of D conditional on S, which I will refer to as the distribution of D given θ and S, can easily be estimated. One method for estimating the distribution of D conditional on, say, j segregating sites would be to assume a value of θ, generate samples and disregard all samples except those with j segregating sites. Examining the remaining samples would allow one to estimate the distribution of D given θ and $S(=j)$. A more efficient method is as follows: for each genealogy generated, place j mutations on the genealogy as described earlier, and calculate the resulting D. This outcome is then weighted by the probability of j mutations on that genealogy, namely:

$$w = \frac{(uT)^j}{j!} e^{-uT},$$

where T is the total length of the genealogy. For example, the probability of $D \geq 1.75$, would be estimated by

$$<P(D \geq 1.75 | S = j)> = \frac{\sum_{i=1}^{m} w_i \chi(D_i \geq 1.75)/m}{\sum_{i=1}^{m} w_i/m},$$

where w_i is the weighting factor given above calculated for the ith genealogy, where D_i is the value of D on the ith genealogy and where $\chi(D_i \geq 1.75)$ is one if D_i is greater than or equal to 1.75, and zero otherwise. The numerator is an estimate of the probability of j segregating sites and $D \geq 1.75$, while the denominator is an estimate of the probability of j segregating sites. Thus, as the number of trees generated becomes large the ratio will converge to the desired conditional probability. To our surprise, we found that the distribution of D given θ and S is more sensitive to θ than the distribution of D given θ (Hudson and Aquadro, in preparation). For example, the probability of $D \geq 1.75$ conditional on 10 segregating sites is 0.053, 0.021, and 0.0079 for $4Nu$ equal to 1.77, 3.53, and 7.07 , respectively. So, again, we would reject neutrality for θ equal to 3.53 and 7.07, but fail to reject for θ equal to 1.77.

A third possibility for generating a distribution of D is one which Dr. Aquadro and I think may be most appropriate for interpreting data when θ is unknown. This method is like the last method described, except that w_i is always set to one. That is, one always places j mutations on the genealogy and gives equal weighting to each genealogy generated. This distribution I will refer to as the random-genealogy distribution of D. The parameter θ does not enter at all in this distribution. If one has no prior information about the value of θ, and one observes 10 segregating sites in a sample, it is perhaps most appropriate to test the null hypothesis that "the sample is the result of 10 mutations being placed on a genealogy randomly drawn from the distribution of genealogies under an equilibrium neutral model." This is the null hypothesis tested by using the random-genealogy distribution of D. The probability of $D \geq 1.75$ when 10 mutations are placed on a random neutral genealogy estimated by simulation is 0.0255, falling just inside the 95% confidence region for D in this distribution.

In contrast, if one knew beforehand what the value of θ is for this genomic region, it would be appropriate to test a null hypothesis of an equilibrium neutral model using the distribution of D given θ, and perhaps the distribution of D given θ and S.

Summarizing, I have described three distributions of D: (1) the distribution of D given θ in which the value of θ is given, but S varies from sample to sample, (2) the distribution of D given θ and S, in which θ and S are given, and (3) the random-genealogy distribution of D, in which a given number of mutations, S, is sprinkled on random genealogies. This third distribution may be the most appropriate for testing neutrality when θ is unknown and is very distinct conceptually from the first two distributions.

TWO-LOCUS AND MANY-LOCUS MODELS

Now I will consider a two-locus neutral model. Specifically, a random union of gametes Wright-Fisher model with constant population size will be considered. This model has the sensible property that the marginal distribution of the gene genealogy of one of the two loci is simply the one locus distribution discussed in the previous section. What is new in the two-locus model is the correlation between the genealogies of the two loci. In fact, one way of generating two-locus genealogies is to generate the genealogy of one locus by the one-locus method, then generate the genealogy of the second locus conditional on the genealogy of the first locus.

To illustrate some of the basic processes involved in two-locus genealogies, I will describe the two-locus history of a sample of size one. In this case, the history is characterized by the time intervals during which the ancestors of the sampled genes reside on the same chromosome. Figure 2 shows an example of the two-locus history of a sample of one gamete. The duration of the first time

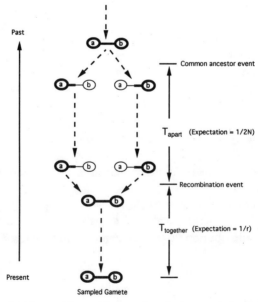

Fig. 2. An example of a two-locus history of a single sampled gamete. The bold-face portions of the chromosomes are direct ancestors of the sampled gamete, the non-bold-face portions are not direct ancestors of the sampled gamete.

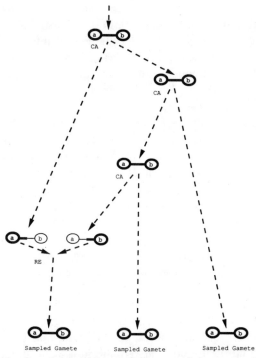

Fig. 3.   An example of a two-locus history of three sampled gametes. As in the last figure, the bold-face portion of the chromosomes are direct ancestors of the sampled gametes. Recombination events are labeled "RE". The occurrence of most recent common ancestors of the sampled alleles is indicated with "CA".

interval during which the ancestors of the sampled genes reside on the same chromosome is exponentially distributed with mean $1/r$, where r is the probability of recombination per generation per gamete in the population. The duration of the second time interval during which the ancestors of the sampled genes are on separate chromosomes is also exponentially distributed, but with mean equal to $2N$ generations. This is the mean time back to the common ancestor of two randomly chosen genes. For larger samples, similar considerations allow one to generate the genealogies of the two loci simultaneously. (See Hudson(1983) for a more detailed description of multilocus genealogies.) Figure 3 shows the two-locus history of a sample of three gametes showing both recombination events and the occurrence of the most recent common ancestors of the various alleles at different times in the history of the sample.

Being able to generate two-locus genealogies obviously allows one to efficiently study two-locus sampling properties. To illustrate, I will describe how to estimate, under an equilibrium two-locus neutral infinite-allele model, the joint distribution of p, q, and D, where p is the frequency of the most common allele at locus 1, q is the frequency of the most common allele at locus 2, and D is the linkage disequilibrium between the two alleles (not to be confused with Tajima's D statistic used earlier).

Again, to make the discussion more concrete, I will consider a specific observation and make use of simulations to help interpret the observation. This time the observation is a real one, taken from a restriction site survey by Aquadro *et al.* (1986) of 48 lines of *D. melanogaster* from the eastern United States. They found eight polymorphic restriction sites in a 13 kb region that included the *Adh* locus. They also determined the ADH electromorph (S or F) of each line. One of the remarkable aspects of their data was the very large and significant linkage disequilibrium observed between a *Bam*HI restriction site polymorphism and the electromorph. These polymorphisms are due to nucleotide polymorphisms at sites about 7.3 kb apart. The frequency of the restriction site was 0.625 in the 48 lines and the frequency of the S electromorph was 0.667 in their lines. The observed number of lines with both the restriction site and the S electromorph was 29 which gives a value of linkage disequilibrium of 0.1875. I will address the question: how likely is this amount of linkage disequilibrium under an equilibrium neutral model, given the observed allele frequencies at each site? This will clearly depend on the unknown recombination rate between these sites, so I will consider several possible rates of recombination.

A brute force approach would be to generate two-locus genealogies, and then following the approach for the one-locus model described above, generate two-locus haplotypes, then tabulate all the results. The problem with this scheme is that one might have to generate thousands of samples to get even a small number of samples with $p=0.625$ and $q=0.667$ which are the samples of interest. A more efficient method is as follows.

For each two-locus genealogy, examine all the branches of the locus 1 genealogy, to find those branches which, if a mutation occurred on them, would result in an allele of frequency p or $1-p$ in the sample. Call this set of branches on the genealogy of locus 1 which lead to np descendents or $n(1-p)$ descendents, the p branches. Call the set of branches on the genealogy of locus 2 that lead to nq or $n(1-q)$ descendents, the q branches. Then, for a given two-locus genealogy, the probability of finding two alleles at locus 1, with one of the alleles at frequency p, is the probability of one or more mutations on one of the p branches of locus 1 and no mutations elsewhere on that genealogy. And the probability of two alleles at locus 2, with one of the alleles at frequency q, is the probability of one or more mutations on one of the q branches of the genealogy of locus 2 and no mutations elsewhere on that genealogy. If the mutation rate is u, and the length of a p branch is t_p, then for a given locus 1 genealogy, the probability of an allele of frequency p due to a mutation on that p branch is

$$(1 - e^{-ut_p})\, e^{-u(T_1 - t_p)},$$

where T_1 is the total length of the locus 1 genealogy. A similar probability can be calculated for any q branch on the locus 2 genealogy. The value of linkage disequilibrium, D, that would result from a mutation on a particular p branch and a mutation on a particular q branch can be determined from the two-locus genealogy. So to estimate the joint probability of p, q, and D, I calculate the following expression for m two-locus genealogies:

$$<P(p,q,D=x)> = \sum_{i=1}^{m} \sum_{\substack{j \in p \text{ branches,} \\ k \in q \text{ branches}}} (1 - e^{-ut_j})e^{-u(T_1-t_j)}(1 - e^{-ut_k})e^{-u(T_2-t_k)}\chi(D_{j,k}=x)/m,$$

where the inner sum is over all pairs (j,k) of p branches and q branches of the ith sample genealogy, where $D_{j,k}$ is the linkage disequilibrium resulting from a mutation on branch j of the locus 1 genealogy and a mutation on branch k of the locus 2 genealogy and where $\chi(D_{j,k}=x)$ is one if $D_{j,k}$ is equal to x and zero otherwise. This is the method employed by Hudson (1985) to estimate the joint distribution of p, q, and D. By replacing $\chi(D_{j,k}=x)$ by one in the expression above, one can estimate the probability of p and q, and also then estimate the probability of $D=x$ given p and q by:

$$<P(D=x|p,q)> = \frac{\displaystyle\sum_{i=1}^{m} \sum_{\substack{j \in p \text{ branches,} \\ k \in q \text{ branches}}} (1 - e^{-ut_j})e^{-u(T_1-t_j)}(1 - e^{-ut_k})e^{-u(T_2-t_k)}\chi(D_{j,k}=x)/m}{\displaystyle\sum_{i=1}^{m} \sum_{\substack{j \in p \text{ branches,} \\ k \in q \text{ branches}}} (1 - e^{-ut_j})e^{-u(T_1-t_j)}(1 - e^{-ut_k})e^{-u(T_2-t_k)}/m}$$

$$\approx \frac{\displaystyle\sum_{i=1}^{m} \sum_{\substack{j \in p \text{ branches,} \\ k \in q \text{ branches}}} ut_j ut_k \chi(D_{j,k}=x)/m}{\displaystyle\sum_{i=1}^{m} \sum_{\substack{j \in p \text{ branches,} \\ k \in q \text{ branches}}} ut_j ut_k/m}$$

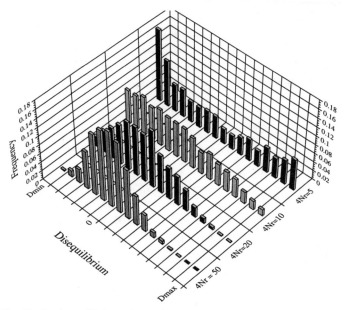

Fig. 4. The distribution of linkage disequilibrium in samples of size 48 conditional on the frequency of most common allele being 0.625 at locus 1 and 0.667 at locus 2.

$$= \frac{\displaystyle\sum_{i=1}^{m} \sum_{\substack{j\in p \text{ branches,} \\ k\in q \text{ branches} \\ q \text{ branches}}} t_j t_k \chi(D_{j,k}=x)/m}{\displaystyle\sum_{i=1}^{m} \sum_{\substack{j\in p \text{ branches,} \\ k\in q \text{ branches}}} t_j t_k/m},$$

where the approximation relies on u, the mutation rate, being sufficiently small that uT is almost always much less than one. Thus as θ becomes small the conditional distribution of D approaches a distribution independent of the precise value of θ.

Figure 4 shows estimates of the probability of $D=x$ given $p=0.625$ and $q=0.667$ obtained by the last expression above, with $m=100,000$. Four different levels of recombination were used, $4Nr=50$, 20, 10, and 5. For this sample size and this p and q, there are 17 possible values of D, and the observed value, 0.1875, is the second largest possible. As can be seen from Fig. 4, such an extreme value of D is very unlikely unless $4Nr$ is 10 or less. For $4Nr=20$ the probability of D greater than or equal to 0.1875 conditional on

$p=0.625$ and $q=0.667$ is approximately 0.002. What is a realistic value of $4Nr$? Although there is great uncertainty, the following calculation can give a rough idea. The effective population size in *D. melanogaster* has been estimated at about 10^6 (Kreitman, 1983). The per base pair per generation recombination rate in females has been estimated at about 1.7×10^{-8} (Chovnick *et al.*, 1977). So $4Nr$ for two sites separated by 7.3 kb would be $(0.5)(4)(10^6)(1.7 \times 10^{-3})(7.3 \times 10^3) = 248$, where the factor of 0.5 is due to the fact that recombination is essentially absent in males. Even if our estimate of $4Nr$ is off by a factor of 10, the observed level of linkage disequilibrium is too high to be compatible with the equilibrium neutral model. Of course, there is a great deal of other evidence that the equilibrium neutral model is incompatible with the observed patterns of genetic variation at this locus. A recent rapid increase in the frequency of the F allele has been suggested and might account for the observed linkage disequilibrium. The involvement of inversions and various forms of selection are also possible. One note of caution: the particular pair of polymorphic sites that I have chosen to examine is one pair out of a set of 36 possible pairs of polymorphic sites in this sample and it was chosen specifically because of its high level of linkage disequilibrium. Nevertheless, these results suggest that the observation is quite unlikely under the neutral equilibrium model if recombination is anywhere near the value obtained above. A proper analysis of the whole data set requires that one examine the data set as a whole without picking out an unusual pair of sites. A more sophisticated approach is necessary to carry out such an analysis.

I will now describe another use of the joint distribution of p, q, and D for estimating the recombination parameter of a neutral model. Consider a many-site model with recombination possible between any of the sites. Let R denote $4Nr$ where r is the recombination rate per generation between the ends of the genomic region being studied. Suppose that a sample of sequences is obtained for this region and that s segregating sites are found. There are $s(s-1)/2$ pairs of polymorphic sites, each pair can be characterized by p, q, D and a distance, x, measured as a fraction of the total length of the region being examined. If R were known, the results described in the previous section could be used to estimate the probability of the observed value of p, q, and D, for each pair of sites. That is, for the ith pair, we could estimate $P(p_i, q_i, D_i | Rx_i)$, Rx_i being the product of $4N$ and the recombination rate between the pair of sites. If all the pairs of sites were statistically independent of each other, the probability of observing a set of p, q, and D's would be

$$L = \prod_{\text{pairs}} P(p_i, q_i, D_i | Rx_i).$$

Now, for a set of tightly linked polymorphic sites, the p, q, and D's are clearly not independent for different pairs. However, I will show below, for one set of parameter values, that an estimator, $<R>$, obtained by maximizing the quantity L defined above is a fairly good estmator of R. Clearly, to obtain this estimate requires estimates of the joint distribution of p, q, and D, which I have shown how to obtain.

To assess the properties of this estimator, $<R>$, where the data are from many tightly linked sites, requires the analysis of many-locus models. Using the coalescent process, one can generate data under many-locus models with recombination. I will not describe the methods here, but note only that the methods are straightforward extensions of the two-locus model described here (Hudson, 1983). Such methods could be easily used to investigate sampling properties of models which are appropriate for the interpretation of DNA sequence data or restriction site polymorphism data.

I have begun to examine the properties of the estimator, $<R>$, using computer simulations and found that it outperforms all other published methods of estimating $4Nr$ using this sort of data. An example of the distribution of this estimator is shown in Fig. 5 where I have also shown the distribution of another estimator based on S_k^2 described by Hudson (1987). The true value of R in this example is 40 and the mean and variance of $<R>$ is 40.4 and 206,

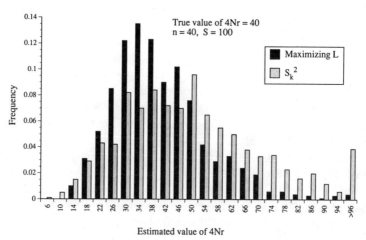

Fig. 5. The distribution of two estimators of $4Nr$ with samples of size 40 and with 100 segregating sites in the sample. The true value of $4Nr$ is this case was 40. The method based on maximizing L is described in the text. The method based on S_k^2 is the method of Hudson (1985).

while the mean and variance of the estimator based on S_k^2 is 49.9 and 498. More detailed descriptions of the properties of this new estimator of R will appear elsewhere.

CONCLUSIONS

The coalescent process provides an extremely efficient method of investigating the distribution of sample statistics under the neutral model. Consideration of the gene genealogies leads to consideration of sample distributions that would not even be considered in the traditional population genetics of alleles and allele frequencies. The random-genealogy distribution of Tajima's D is an example. For multilocus models, simulations may be the only way to obtain detailed information about the distributions of a great variety of sample statistics, including test statistics and estimators that will be applied to the interpretation of DNA sequence data.

SUMMARY

A method of generating single locus gene genealogies is described. The method is applied to examining the distribution of Tajima's test statistic, D. A distribution of D is described for use when the mutation parameter $4Nu$ is unknown. This distribution, referred to as the random-genealogy distribution, is generated without using any value of $4Nu$. Two-locus and multilocus genealogies are briefly considered. To illustrate their use in estimating two-locus sampling properties, a method is described for estimating the joint distribution of allele frequencies and linkage disequilibrium. A new method of estimating the recombination parameter of the neutral model is described which uses the two-locus sampling distribution. The properties of this estimator are described for one set of parameter values to demonstrate the utility of multi-locus genealogy simulations. This new estimator is considerably better than an earlier published method.

Acknowledgments
I am grateful to Charles Aquadro for comments on this work which was supported by grant GM42447 from the National Institutes of Health, USA.

REFERENCES

Aquadro, C.F., M.M. Deese, C.H. Bland, C.H. Langley, and C.C. Laurie-Ahlberg, 1986.

Genetics **114**: 1165-1190.

Chovnick, A., W. Gelbart, and M. McCarron, 1977. *Cell* **11**: 1-10.

Hudson, R.R., 1983. *Theor. Popul. Biol.* **23**: 183-201.

Hudson, R.R., 1985. *Genetics* **109**: 611-631.

Hudson, R.R., 1987. *Genet. Res. Camb.* **50**: 245-250.

Hudson, R.R., 1990. In *Oxford Surveys in Evolutionary Biology*, Vol. 7, edited by J. Antonovics and D. Futuyma, pp. 1-44. Oxford University Press, Oxford.

Kreitman, M., 1983. *Nature* **304**: 412-417.

Tajima, F., 1989. *Genetics* **123**: 585-595.

Measurement of DNA Polymorphism

FUMIO TAJIMA

Department of Population Genetics, National Institute of Genetics, Mishima 411, Japan

A large amount of DNA polymorphism is maintained in natural populations. The amount of DNA polymorphism is determined by many factors: mutation, recombination, migration, natural selection, change in population size, population subdivision, random genetic drift, historical accident, and others. In order to identify the mechanism of the maintenance of DNA polymorphism, we must first know the amount of DNA polymorphism.

In the past when the only data available were from electrophoretic mobility of protein and serological typing of blood group, the average heterozygosity or the number of alleles was used as a measure of the amount of genetic variation in a population (Lewontin, 1974; Nei, 1975). Now we can study the genetic variation at the DNA level (Kimura, 1983; Nei, 1987).

There are two different quantities for measuring the amount of genetic variation at the DNA level, *i.e.*, the average number of pairwise nucleotide differences and the number of segregating (polymorphic) sites among a sample of DNA sequences. In this paper I shall discuss the differences between the two quantities, and examine the statistical properties under several models, including natural selection, migration, and change in population size.

37

HOW TO MEASURE THE AMOUNT OF DNA POLYMORPHISM

In order to show how to measure the amount of DNA polymorphism, let us consider the following four DNA sequences with a length of 20 nucleotides:

Sequence 1	ACTGG**C**TAAGCGC**AT**ACTAG
Sequence 2	ACTGG**CG**AAGCCCATG**C**TAG
Sequence 3	AC**C**GG**TG**AAGT**C**CATG**C**TTG
Sequence 4	AC**C**GG**CG**AAGCCCATG**C**TAG.

The number of segregating sites (S) is the number of sites which are occupied by at least two different nucleotides. As shown by bold letters, there are seven such sites, so that $S=7$. The average number of pairwise nucleotide differences (k) among DNA sequences is defined as

$$k=\sum_{i=1}^{n-1}\sum_{j=i+1}^{n}k_{ij}\Big/\binom{n}{2},$$

where k_{ij} is the number of nucleotide differences between sequences i and j, n is the number of DNA sequences sampled from a population and $\binom{n}{2}=n(n-1)/2$ is the total number of sequence comparisons. In the above example, three nucleotides are different between sequences 1 and 2, so that $k_{12}=3$. In the same way, we have $k_{13}=7$, $k_{14}=4$, $k_{23}=4$, $k_{24}=1$, and $k_{34}=3$. Then, the average number of nucleotide differences (k) among DNA sequences is given by $k=(k_{12}+k_{13}+k_{14}+k_{23}+k_{24}+k_{34})/6=(3+7+4+4+1+3)/6=3.667$. When the number of DNA sequences is large, this method is time-consuming. In such a case, it is recommended to utilize heterozygosity. Heterozygosity at site i is defined as

$$1-\sum_{j=1}^{4}p_{j}^{2},$$

where p_j is the relative frequency of nucleotide j ($j=1, 2, 3,$ and 4 correspond to A, T, G, and C), and the unbiased estimate is given by

$$h_i=n(1-\sum_{j=1}^{4}x_{ij}^{2})/(n-1),$$

where x_{ij} is the observed frequency of nucleotide j at site i and n is the number of DNA sequences. Obviously $h_i=0$ at a monomorphic site, so that it is sufficient to consider only polymorphic sites. At site 3, $x_{31}=0$, $x_{32}=0.5$, $x_{33}=0$, and $x_{34}=0.5$ so that $h_3=4(1-0^2-0.5^2-0^2-0.5^2)/3=0.667$. In the same way, we have $h_6=h_7=h_{11}=h_{12}=h_{16}=h_{19}=0.5$. Then, the average number of nucleotide differences (k) is given by

$$k = \sum_{i=1}^{m} h_i,$$

where m is the number of nucleotide sites in the DNA sequence (Nei and Miller, 1990). In this example, we have $k = 0.667 + 0.5 + 0.5 + 0.5 + 0.5 + 0.5 + 0.5 = 3.667$, which is the same as that obtained previously. Which method is better depends on the number of segregating sites (S) and the number of DNA sequences (n).

Both S and k depend on the length of DNA sequence (m), and the amount of DNA polymorphism per site can be used instead, which is obtained by dividing S or k by m. In this example, the number of segregating sites per site is $S/m = 7/20 = 0.35$ and the average number of nucleotide differences per site is $k/m = 3.667/20 = 0.1833$. As might be clear from the above equation, the latter can be called the average heterozygosity per site.

MEASUREMENT OF DNA POLYMORPHISM UNDER THE NEUTRAL THEORY

Here, we examine statistical properties of the above two measures of DNA polymorphism under the neutral theory (Kimura, 1968, 1983). We consider a random mating and equilibrium population with N diploid individuals.

1. Number of Segregating Sites

Watterson (1975) showed that when mutations are selectively neutral and when there is no recombination, the mean and variance of the number of segregating sites (S) among a sample of n DNA sequences are given by

$$E(S) = a_1 M \text{ and } V(S) = a_1 M + a_2 M^2, \tag{1}$$

respectively, where

$$a_1 = \sum_{i=1}^{n-1} (1/i), \ a_2 = \sum_{i=1}^{n-1} (1/i^2), \tag{2}$$

$M = 4Nv$, N is the effective population size, and v is the mutation rate per DNA sequence per generation. These equations indicate that S depends on n, so that S is not a good measure of the amount of DNA polymorphism. When the neutral theory is correct, however, S/a_1 can be used as an estimate of M, namely,

$$\hat{M} = S/a_1. \tag{3}$$

The variance of $\hat{M}$ can be estimated by

$$V(\hat{M}) = \frac{a_1^2 S + a_2 S^2}{a_1^2(a_1^2 + a_2)}, \tag{4}$$

which can be obtained by using Eq. (1) and $E(S^2) = a_1 M + (a_1^2 + a_2)M^2$ (see Tajima, 1989b). In the above example, $S = 7$, $a_1 = 1 + 1/2 + 1/3 = 1.833$, and $a_2 = 1 + 1/4 + 1/9 = 1.361$, so that we obtain $\hat{M} = 3.818$ and $V(\hat{M}) = 5.684$.

When n is large (say, $n > 10$), computations of a_1 and a_2 are cumbersome. In this case the following approximations are recommended:

$$a_1 \approx \gamma + \log_e(n - 0.5) + \frac{1}{24n(n-1)} \tag{5}$$

$$\approx \gamma + \log_e(n - 0.5), \tag{6}$$

$$a_2 \approx \pi^2/6 - 1/(n - 0.5), \tag{7}$$

where $\gamma = 0.5772156649\cdots$ (Euler's constant) and $\pi^2/6 = 1.644934066\cdots$. In the above example, the sample size is four ($n = 4$), so that we have $a_1 \approx 1.833$ from Eq. (5), $a_1 \approx 1.830$ from Eq. (6), and $a_2 \approx 1.359$ from Eq. (7), which are very close to true values ($a_1 = 1.8333\cdots$ and $a_2 = 1.3611\cdots$) even when n is as small as 4. The absolute errors caused by the approximations are shown in Fig. 1, which indicates that if we want to keep the absolute error less than 10^{-3}, then Eqs. (5), (6), and (7) can be used for $3 \leq n \leq 6$, $n \geq 7$, and $n \geq 5$, respectively. In the case where the absolute error less than 10^{-4} is required, Eqs. (5), (6), and (7) can be used for $5 \leq n \leq 20$, $n \geq 21$, and $n \geq 10$, respectively.

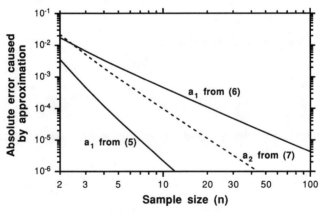

Fig. 1. Absolute errors of a_1 $\{= \sum_{i=1}^{n-1}(1/i)\}$ and a_2 $\{= \sum_{i=1}^{n-1}(1/i^2)\}$ caused by approximations.

Since we have assumed that there is no recombination between sites, Eq. (4) gives an overestimate if there is recombination. In this case, a lower bound of $V(\hat{M})$ can be obtained by assuming linkage equilibrium. It is given by

$$V_{\min}(\hat{M}) = S/a_1^2. \tag{8}$$

In the above example we have $V_{\min}(\hat{M}) = 2.083$, which is substantially smaller than that for the case of no recombination.

2. Average Number of Nucleotide Differences

Tajima (1983) showed that when mutants are selectively neutral and when there is no recombination between sites, the mean and variance of the average number of pairwise nucleotide differences (k) among a sample of n DNA sequences are given by

$$E(k) = M \text{ and } V(k) = b_1 M + b_2 M^2, \tag{9}$$

respectively, where b_1 and b_2 are given by

$$b_1 = \frac{n+1}{3(n-1)} \text{ and } b_2 = \frac{2(n^2+n+3)}{9n(n-1)}. \tag{10}$$

Thus, k not only has a clear biological meaning, but also gives the estimate of M directly. This measure, however, has a drawback. That is, the variance of k is not small even when a large number of DNA sequences is studied. In fact, as n increases, the variance approaches

$$V_{st}(k) = \frac{1}{3}M + \frac{2}{9}M^2, \tag{11}$$

which is called the stochastic variance (Tajima, 1983). The sampling variance is given by

$$V_s(k) = V(k) - V_{st}(k) = \frac{2}{3(n-1)}M + \frac{2(2n+3)}{9n(n-1)}M^2 \tag{12}$$

(Tajima, 1983), which approaches zero as n increases.

The total variance, stochastic variance, and sampling variance of k can be estimated by

$$\hat{V}(k) = \frac{3n(n+1)k + 2(n^2+n+3)k^2}{11n^2 - 7n + 6}, \tag{13}$$

$$\hat{V}_{st}(k) = \frac{(3n^2 - 3n + 2)k + 2n(n-1)k^2}{11n^2 - 7n + 6}, \tag{14}$$

$$\hat{V}_s(k) = \frac{2(3n-1)k + 2(2n+3)k^2}{11n^2 - 7n + 6}, \tag{15}$$

respectively. These formulae can be obtained from Eqs. (9), (10), (11), and (12). In the previous example, $k = 3.667$ and $n = 4$, so that we have $\hat{V}(k) = 5.444$, $\hat{V}_{st}(k) = 3.000$, and $\hat{V}_s(k) = 2.444$.

As in the case of the number of segregating sites, Eqs. (13), (14), and (15) give overestimates if there is recombination. The minimum estimates can be obtained when there is no linkage disequilibrium, and can be given by

$$\hat{V}_{min}(k) = \frac{n+1}{3(n-1)}k, \tag{16}$$

$$\hat{V}_{st,min}(k) = \frac{1}{3}k, \tag{17}$$

$$\hat{V}_{s,min}(k) = \frac{2}{3(n-1)}k. \tag{18}$$

In the above example, we have $\hat{V}_{min}(k) = 2.037$, $\hat{V}_{st,min}(k) = 1.222$, and $\hat{V}_{s,min}(k) = 0.815$, which are substantially smaller than those for the case of no recombination.

3.   Test of the Neutral Mutation Hypothesis

Hudson *et al.* (1987) developed a statistical method for testing the neutral mutation hypothesis. Their method requires not only data on the amount of DNA polymorphism but also data on the evolutionary distance between two species at the DNA level. On the other hand, only data on the amount of DNA polymorphism is required in Tajima's (1989b) method, which is summarized here.

As mentioned in the above, under the neutral theory the expectation of S/a_1 and that of k are both M. Therefore, if S/a_1 significantly deviates from k, then we can reject the neutral mutation hypothesis. Tajima (1989b) proposed the following statistic:

$$D = \frac{k - S/a_1}{\sqrt{e_1 S + e_2 S(S-1)}}, \tag{19}$$

where

$$e_1 = \frac{b_1 - 1/a_1}{a_1},$$

$$e_2 = \frac{b_2 - (n+2)/(a_1 n) + a_2/a_1^2}{a_1^2 + a_2}.$$

Then, he showed that D follows a beta distribution with mean 0 and variance 1, approximately. If D is significantly different from 0, then we can reject the neutral mutation hypothesis. Furthermore, if D is significantly smaller than 0, then it suggests that purifying selection is operating. In contrast, if D is significantly larger than 0, it suggests that balancing selection such as overdominant selection is operating. The confidence limit of D for a given n can be obtained from Table 2 of Tajima (1989b). It should be noted that this test is conservative since it depends on the assumption that there is no recombination between sites.

In the previous example, we have $k=3.667$, $S=7$, $n=4$, $a_1=1.833$, $a_2=1.361$, $b_1=5/9=0.556$, and $b_2=23/54=0.426$, so that $e_1=0.00551$ and $e_2=0.00269$. Then, we finally have $D=-0.389$. We can see from Table 2 of Tajima (1989b) that the 90% confidence limit is $-0.876<D<2.081$ for $n=4$, so that the neutral mutation hypothesis cannot be rejected.

Tajima (1989b) applied this test to restriction enzyme data of *Drosophila melanogaster*, and concluded that only data for large deletions and insertions, which might be caused by transposable elements, do not support the neutral mutation hypothesis but are consistent with the deleterious selection model.

EFFECT OF MIGRATION ON DNA POLYMORPHISM

When a population is subdivided, a large amount of DNA polymorphism can be maintained in the population if the migration rates among subpopulations are small. Here, we examine the effect of migration or population subdivision on the amount of DNA polymorphism under the assumption that mutations are selectively neutral.

1. Two-subpopulation Model

Assume that a population is subdivided to two subpopulations, say, subpopulations 1 and 2. Denote the effective sizes of subpopulations 1 and 2 by N_1 and N_2, the migration rate from subpopulation 1 to subpopulation 2 (*i.e.*, the probability that a DNA sequence in subpopulation 2 comes from subpopulation 1 in the immediately previous generation) by m_2, that from subpopulation 2 to subpopulation 1 by m_1, and the mutation rate per DNA sequence per generation by v. Assuming that m_1 and m_2 are so small that their higher order terms can be ignored, Tajima (1989a) showed that the expected number of segregating sites among a sample of n DNA sequences, of which i and j ($=n-i$) sequences are randomly sampled from subpopulations 1 and 2, respectively, can be given by

$$S(i,j)=\frac{nM_1M_2+a_{ij}M_1+b_{ij}M_2}{j(j-1+Q_2)M_1+i(i-1+Q_1)M_2}, \tag{20}$$

where

$$a_{ij}=j\{(j-1)S(i,j-1)+Q_2S(i+1,j-1)\},$$

$$b_{ij}=i\{(i-1)S(i-1,j)+Q_1S(i-1,j+1)\},$$

$M_1=4N_1v$, $M_2=4N_2v$, $Q_1=4N_1m_1$, and $Q_2=4N_2m_2$. Since $S(1,0)=S(0,1)=0$, we can obtain $S(i,j)$ by using Eq. (20) repeatedly.

The expected number of segregating sites in n DNA sequences randomly sampled from the entire population can be given by

$$S(n)=\sum_{i=0}^{n}P(i,n-i)S(i,n-i) \tag{21}$$

(Tajima, 1989a), in which $P(i,n-i)$ is given by

$$P(i,n-i)=\frac{n!}{i!(n-i)!}\frac{M_1^iM_2^{n-i}}{(M_1+M_2)^n}.$$

For example, when $n=2$, from Eq. (20) we have

$$S(2,0)=\frac{M_1+Q_1S(1,1)}{1+Q_1}, \tag{22a}$$

$$S(1,1)=\frac{2M_1M_2+Q_2M_1S(2,0)+Q_1M_2S(0,2)}{Q_2M_1+Q_1M_2}, \tag{22b}$$

$$S(0,2)=\frac{M_2+Q_2S(1,1)}{1+Q_2}, \tag{22c}$$

which are equivalent to Eq. (9) of Strobeck (1987) and Eq. (13) of Slatkin (1987). Solving Eq. (22), we obtain

$$S(2,0)=\{Q_2(1+Q_2)M_1^2+Q_1(3+2Q_2)M_1M_2+Q_1^2M_2^2\}/A, \tag{23a}$$

$$S(1,1)=\{Q_2(1+Q_2)M_1^2+2(1+Q_1)(1+Q_2)M_1M_2+Q_1(1+Q_1)M_2^2\}/A, \tag{23b}$$

$$S(0,2)=\{Q_2^2M_1^2+Q_2(3+2Q_1)M_1M_2+Q_1(1+Q_1)M_2^2\}/A, \tag{23c}$$

where $A=Q_2(1+Q_2)M_1+Q_1(1+Q_1)M_2$. These equations give the expectation of the average number of pairwise nucleotide differences, since it is equal to the expected number of segregating sites among a sample of two DNA sequences, as mentioned earlier. Therefore, from Eq. (21) the expectation of the average number of nucleotide differences among DNA sequences randomly sampled from the entire population can be given by

TABLE 1

The Expectation of the Average Number of Nucleotide Differences, $E(k)$, and the Expectation of the Estimate of M, $E(\hat{M}) = E(S)/a_1$, Obtained from the Number of Segregating Sites among n DNA Sequences Randomly Sampled Either from Subpopulation 1 or from the Entire Population, Where $M_1 = M_2 = 1$ and $Q_1 = Q_2 = 1$ Were Assumed

Sample size	Subpopulation 1		Entire population	
(n)	$E(\hat{M})$	$E(k)$	$E(\hat{M})$	$E(k)$
2	2.000	2.000	2.500	2.500
6	1.972	2.000	2.459	2.500
10	1.938	2.000	2.413	2.500
20	1.885	2.000	2.355	2.500
50	1.813	2.000	2.294	2.500

TABLE II

The Expectation of the Average Number of Nucleotide Differences, $E(k)$, and the Expectation of the Estimate of M, $E(\hat{M}) = E(S)/a_1$, Obtained from the Number of Segregating Sites among 50 DNA Sequences Randomly Sampled Either from Subpopulation 1 or from the Entire Population, Where $M_1 = M_2 = 1$ and $Q_1 = Q_2$ Were Assumed

Q_1	Subpopulation 1		Entire population	
	$E(\hat{M})$	$E(k)$	$E(\hat{M})$	$E(k)$
0.01	1.992	2.000	24.453	52.000
0.1	1.937	2.000	4.349	7.000
1	1.813	2.000	2.294	2.500
10	1.898	2.000	2.042	2.050
100	1.980	2.000	2.005	2.005

$$E(k) = S(2) = P(0,2)S(0,2) + P(1,1)S(1,1) + P(2,0)S(2,0), \qquad (24)$$

where $P(0,2) = M_2^2/(M_1 + M_2)^2$, $P(1,1) = 2M_1M_2/(M_1 + M_2)^2$, and $P(2,0) = M_1^2/(M_1 + M_2)^2$.

Using these equations, let us examine the effect of population subdivision or migration on the measures of DNA polymorphism. First, we consider the case where the size of two subpopulations is the same and the migration rate from subpopulation 2 to subpopulation 1 is the same as that from subpopulation 1 to subpopulation 2, i.e., $M_1 = M_2$ and $Q_1 = Q_2$. Table I shows the case where $M_1 = M_2 = 1$ and $Q_1 = Q_2 = 1$, and indicates that the expectation of the average number of nucleotide differences, $E(k)$, does not depend on n, whereas the expectation of the estimate of M obtained from the number of segregating sites, $E(\hat{M})$, depends on n. This implies that the latter is not a good measure. In other words, the number of segregating sites cannot be used for estimating

TABLE III

The Expectation of the Average Number of Nucleotide Differences, $E(k)$, and the Expectation of the Estimate of M, $E(\hat{M})=E(S)/a_1$, Obtained from the Number of Segregating Sites among n DNA Sequences Randomly Sampled Either from Subpopulation 1 or 2 or from the Entire Population, Where $M_1=0.1$, $M_2=1.9$, and $Q_1=Q_2$ Were Assumed

Q_1	n	Subpopulation 1		Subpopulation 2		Entire population	
		$E(\hat{M})$	$E(k)$	$E(\hat{M})$	$E(k)$	$E(\hat{M})$	$E(k)$
0.01	2	0.307	0.307	2.089	2.089	3.881	3.881
	10	0.306	0.307	2.089	2.089	4.744	3.881
	50	0.305	0.307	2.088	2.089	5.910	3.881
0.1	2	0.445	0.445	2.082	2.082	2.250	2.250
	10	0.440	0.445	2.079	2.082	2.318	2.250
	50	0.428	0.445	2.072	2.082	2.397	2.250
1	2	1.145	1.145	2.045	2.045	2.057	2.057
	10	1.085	1.145	2.041	2.045	2.058	2.057
	50	0.966	1.145	2.031	2.045	2.056	2.057
10	2	1.845	1.845	2.008	2.008	2.009	2.009
	10	1.770	1.845	2.008	2.008	2.009	2.009
	50	1.587	1.845	2.006	2.008	2.008	2.009
100	2	1.983	1.983	2.001	2.001	2.001	2.001
	10	1.970	1.983	2.001	2.001	2.001	2.001
	50	1.916	1.983	2.001	2.001	2.001	2.001

M in a subdivided population even when the neutral theory is correct. Table II shows the effect of migration rate on the estimates of the amount of DNA polymorphism, where $n=50$ was assumed. It is clear that $E(k)$ for a subpopulation does not depend on the migration rate, as indicated by Li (1976) (also see Slatkin, 1987; Strobeck, 1987), whereas $E(\hat{M})$ depends on the migration rate. In the case where the migration rate is small ($Q_1 < 1$), a large difference between $E(\hat{M})$ and $E(k)$ can be seen when DNA sequences are sampled from the entire population.

Table III shows another example, where the effective size of subpopulation 2 is assumed to be 19 times larger than that of subpopulation 1 ($M_2 = 19M_1$). From this table we can see that $E(k)$ is virtually independent of n, but that there is some difference between $E(\hat{M})$ and $E(k)$. For more complicated cases, see Tajima (1989a).

2. Circular Stepping-stone Model

In this model the subpopulations are arranged in a circle and migration is allowed only between adjacent subpopulations, as shown in Fig. 2. We

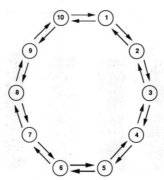

Fig. 2.   Circular stepping-stone model.

assume that the population consists of n subpopulations with N diploid individuals, so that the total population has N_T $(=nN)$ diploid individuals. The mutation rate per DNA sequence per generation is denoted by v, and the migration rate from the j-th subpopulation to the i-th subpopulation is denoted by $m(i,j)$, which is the probability that a DNA sequence in the i-th subpopulation came from the j-th subpopulation in the immediately previous generation. We measure the amount of DNA polymorphism by the expected number of nucleotide differences between two randomly chosen DNA sequences, one from the i-th subpopulation and the other from the j-th subpopulation, and denote it by $K(i,j)$. In an equilibrium population Strobeck (1987) obtained the following relationships:

$$\{1+4N\sum_{j\neq i}m(i,j)\}K(i,i)=4Nv+4N\sum_{j\neq i}m(i,j)K(i,j), \tag{25a}$$

$$2N\{\sum_{k\neq i}m(i,k)+\sum_{j\neq k}m(j,k)\}K(i,j)$$

$$=4Nv+2N\{\sum_{k\neq i}m(i,k)K(j,k)+\sum_{k\neq j}m(j,k)K(i,k)\}, \tag{25b}$$

where $i\neq j$.

Here, we consider only the case where the migration matrix is symmetric, i.e., $m(i,j)=m(j,i)$, and is irreducible so that there are no isolated subsets of subpopulations (see Slatkin, 1987). Slatkin (1987) and Strobeck (1987) have shown that the average of $K(i,i)$ is $4N_Tv$, which is independent of the migration rate. Each $K(i,i)$, however, depends on the migration rate. Figure 3 shows one example, where the number of subpopulations is 10 and the migration rate is the same for all adjacent subpopulations except for subpopulations 5 and 6. Namely, $4Nm(1,10)=4Nm(10,1)=1$ and $4Nm(i,i+1)=4Nm(i+1,i)=1$ for $i\neq5$,

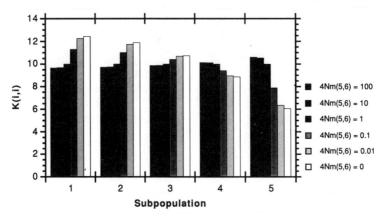

Fig. 3. Expected number of nucleotide differences, $K(i,i)$, between two DNA sequences randomly chosen from the i-th subpopulation, measured in units of $4Nv$, under the circular stepping-stone model. The number of subpopulations is 10. The migration rate between the adjacent subpopulation is 1, except that between subpopulations 5 and 6.

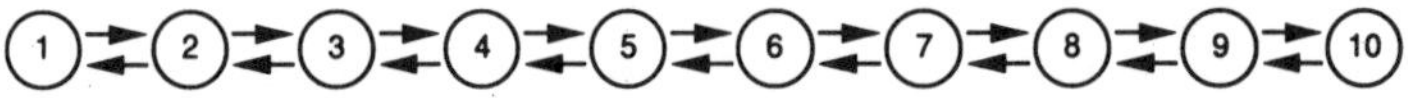

Fig. 4. Linear stepping-stone model.

are assumed, and several values of $4Nm(5,6)=4Nm(6,5)$ are used. In this figure only $K(i,i)$'s for $i\leq5$ are given since $K(i,i)=K(11-i,11-i)$, and are measured in units of $4Nv$. We can see from this figure that the smaller the value of $4Nm(5,6)$ and $4Nm(6,5)$ is, the smaller are $K(5,5)$ and $K(6,6)$. This indicates that the expected number of nucleotide differences between two DNA sequences randomly sampled from the subpopulation with low migration rate is small. This figure also indicates that the amount of DNA polymorphism in the subpopulation with high migration rate tends to be large.

3. Linear Stepping-stone Model

In this model the subpopulations are arranged in a line with two ends, as shown in Fig. 4. This model can be obtained from the circular stepping-stone model directly. Namely, if there is only one barrier between the adjacent subpopulations in the circular model, then the circular stepping-stone model changes into the linear model. When the number of subpopulations is 10 as shown in Fig. 4, the following relationships can be obtained (Tajima, 1990b):

$$4Nv < K(1,1) < K(2,2) \text{ and } 4Nv < K(10,10) < K(9,9).$$

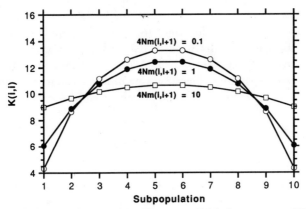

Fig. 5. Expected number of nucleotide differences, $K(i,i)$, between two DNA sequences randomly chosen from the i-th subpopulation, measured in units of $4Nv$, under the linear stepping-stone model. The number of subpopulations is 10.

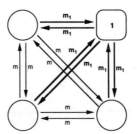

Fig. 6. Finite island model.

This indicates that the expected number of nucleotide differences between two DNA sequences randomly chosen from the same subpopulation at the end is smaller than that from the same subpopulation adjacent to the end.

Numerical examples are shown in Fig. 5, where the number of subpopulations is 10 as in the case of Fig. 4. In this figure the expected number of nucleotide differences between two DNA sequences randomly chosen from the i-th subpopulation, $K(i,i)$, was computed using Eq. (25a) and (25b), and is measured in units of $4Nv$. It is clear from this figure that marginal subpopulations have lower level of DNA polymorphism than central subpopulations.

4. *Finite Island Model*

In this model the migration rates from one subpopulation to any of the other $n-1$ subpopulations have been assumed to be the same. In order to examine the effect of migration on DNA polymorphism in a subpopulation, however, we assume that the migration rate of a particular subpopulation, say subpopulation 1, is different from the others, as shown in Fig. 6. Namely, we assume

$$m(1,i)=m(i,1)=m_1 \text{ and } m(i,j)=m,$$

where $i \neq 1$, $j \neq 1$, and $i \neq j$.

Using Eq. (25a) and (25b) directly, we can obtain the expected number, $K(i,j)$, of nucleotide differences between the two DNA sequences, one from subpopulation i and the other from subpopulation j, which is

$$K(1,1)=\left\{n+\frac{(n-1)(n-2)(4Nm_1-4Nm)}{a}\right\}4Nv, \tag{26a}$$

$$K(i,i)=\left\{n+\frac{(n-2)(4Nm-4Nm_1)}{a}\right\}4Nv, \tag{26b}$$

$$K(1,i)=K(1,1)+\frac{K(1,1)-4Nv}{4(n-1)Nm_1}, \tag{26c}$$

$$K(i,j)=\left\{1+\frac{1}{4(n-1)Nm+4Nm_1}\right\}K(i,i) \tag{26d}$$

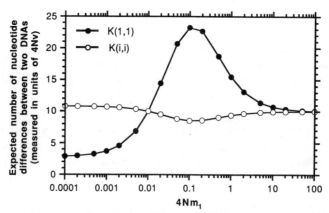

Fig. 7. Expected number of nucleotide differences, $K(i,i)$, between two DNA sequences randomly chosen from the i-th subpopulation, measured in units of $4Nv$, under the finite island model. The number of subpopulations is 10 and $4Nm=0.01$ is assumed.

(Tajima, 1990b), where $a = 16nN^2m_1^2 + 16n(n-1)N^2m_1m + 8Nm_1 + 4nNm$. These equations indicate that $K(1,1)$ is smaller than $4nNv \ (= 4N_Tv)$ and $K(i,i)$ is larger than $4N_Tv$ if m_1 is smaller than m. On the other hand, $K(1,1) > 4N_Tv > K(i,i)$ if $m_1 > m$. This means that the expected amount of DNA polymorphism in the subpopulation with lower migration rate is smaller than that of higher migration rate. Incidentally, the average of $K(i,i)$'s for all i is $\{K(1,1) + (n-1)K(i,i)\}/n = 4N_Tv$, as expected from the studies by Slatkin (1987) and Strobeck (1987).

Numerical examples are shown in Fig. 7, where $4Nm = 0.01$ is assumed. Although the relationship between the expected amount of DNA polymorphism and the migration rate is complicated, we can see that the expected amount of DNA polymorphism in the subpopulation with lower migration rate is always smaller than that of higher migration rate.

EFFECT OF CHANGE IN POPULATION SIZE ON DNA POLYMORPHISM

We have assumed that the size of population is constant. The size of population, however, often changes drastically. Here, we examine the effect of change in population size on the number of segregating sites and the average number of nucleotide differences.

1.   Theory

Let N_t be the effective population size in the t-th generation, and denote by $S_n(t)$ the expected number of segregating sites among n DNA sequences randomly sampled from the population in the t-th generation. It is worth noting that the expectation of the average number of pairwise nucleotide differences is equal to $S_2(t)$.

Assuming that the population size is constant ($N_t = N$ for $t > 0$), Tajima (1989c) has shown that $S_n(t)$ is given by

$$S_n(t) = d_{n,1} + \sum_{i=2}^{n} d_{n,i}\exp(-c_i t), \tag{27}$$

where

$$c_i = \frac{n(n-1)}{4N}, \tag{28a}$$

$$d_{n,1} = d_{n-1,1} + \frac{4Nv}{n-1}, \tag{28b}$$

$$d_{n,i} = \frac{n(n-1)}{(n-i)(n+i-1)}b_{n-1,i} \quad \text{for} \quad 1 < i < n, \tag{28c}$$

$$d_{n,n} = S_n(0) - \sum_{i=1}^{n-1} b_{n,i}. \tag{28d}$$

Because of $S_1(t)=0$, we have $d_{1,1}=0$. From Eq. (28b) we have $d_{n,1}=a_1 M$, where a_1 is given by Eq. (2) and $M=4Nv$ as before. Using Eq. (28) repeatedly, $d_{n,i}$ can be obtained. Then, $S_n(t)$ can be obtained from Eq. (27). For example, when $n=2$, we have

$$S_2(t) = M + \{S_2(0) - M\} \exp\left(-\frac{t}{2N}\right),$$

which is identical with the formula obtained by Li (1977) using a different method. When $n=3$, we have

$$S_3(t) = \frac{3}{2} M + \frac{3}{2}\{S_2(0) - M\} \exp\left(-\frac{t}{2N}\right) + \{S_3(0) - \frac{3}{2} S_2(0)\} \exp\left(-\frac{3}{2N} t\right).$$

When the population is in equilibrium at time 0, we can simplify Eq. (27). Since $S_n(0)=a_1 M_0$, where $M_0=4N_0v$, Eq. (27) becomes

$$S_n(t) = a_1 M + (M_0 - M) \sum_{i=1}^{[n/2]} g_{n,i} \exp(-c_{2i}t) \tag{29}$$

(Tajima, 1989c), where $[n/2]$ is the largest integer that is not greater than $n/2$, and $g_{n,i}$ is given by

$$g_{n,i} = \frac{(n-1)!n!(4i-1)}{(n-2i)!(n+2i-1)!(2i-1)}. \tag{30}$$

For example, when $n=2$, we have

$$S_2(t) = M + (M_0 - M) \exp\left(-\frac{t}{2N}\right).$$

When $n=3$, we have

$$S_3(t) = \frac{3}{2} M + \frac{3}{2}(M_0 - M) \exp\left(-\frac{t}{2N}\right).$$

2. Numerical Examples

We examine three cases, as shown in Fig. 8.

Figure 8a is the case where the size of population suddenly decreases at time 0. Figure 9 shows numerical results obtained from Eq. (29), where $M_0=1$ and $M=0.01$ were assumed (i.e., the size of population suddenly becomes one hundredth at time 0) and $n=2$, 10, and 50 were used. From this figure we can see that the number of segregating sites ($n=10$ and 50) declines more rapidly than the average number of nucleotide differences ($n=2$), and that the

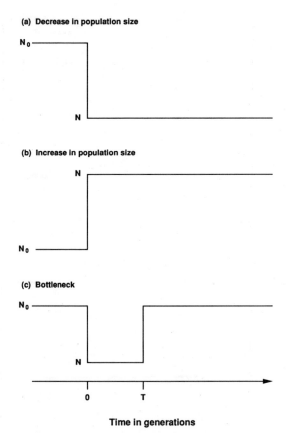

Fig. 8. Models of change in population size.

larger is the sample size, the more quickly the number of segregating sites decreases.

Figure 8b is the case where the size of population suddenly increases at time 0. Figure 10 shows the results obtained from Eq. (29), where $M_0 = 0$ and $M = 1$ were assumed (*i.e.*, until time 0 the size of population is so small that there is no genetic variation, but the size of population becomes large afterwards). It is clear from this figure that the number of segregating sites ($n = 10$ and 50) increases more rapidly than the average number of nucleotide differences ($n = 2$), and that the larger is the sample size, the more quickly the number of segregating sites increases.

Figure 8c shows the case where the size of population becomes small at

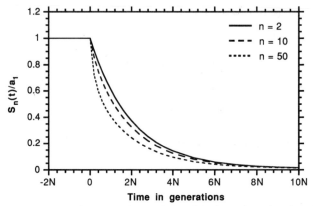

Fig. 9. Expected amount of DNA polymorphism, $S_n(t)/a_1$, when the size of population suddenly decreases at time 0 as shown in Fig. 8a. $M_0 = 1$ and $M = 0.01$ are assumed, and $n = 2$, 10, and 50 are used.

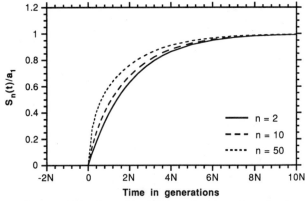

Fig. 10. Expected amount of DNA polymorphism, $S_n(t)/a_1$, when the size of population suddenly increases at time 0 as shown in Fig. 8b. $M_0 = 0$ and $M = 1$ are assumed, and $n = 2$, 10, and 50 are used.

time 0, but the population recovers the original size T generations later. Figure 11 shows the results obtained from Eqs. (27) and (29), where $M_0 = 1$, $M = 0.01$, and $T = 2N$ were assumed. We can see from Fig. 11(a, b) that larger reduction of $S_n(t)/a_1$ is observed when n is larger, but the bottleneck effect continues longer in the case where n is smaller. Therefore, we can conclude that the average number of nucleotide differences is affected by the bottleneck of population size more strongly than is the number of segregating sites.

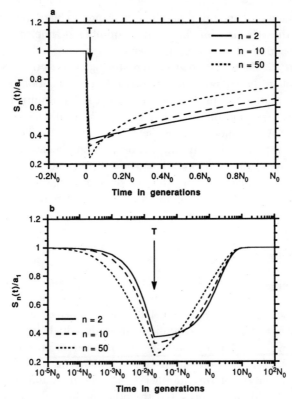

Fig. 11. Expected amount of DNA polymorphism, $S_n(t)/a_1$, when the size of population becomes small at time 0 but the population recovers the original size T generations later, as shown in Fig. 8c. $M_0 = 1$, $M = 0.01$, and $T = 2N$ are assumed, and $n = 2$, 10, and 50 are used.

EFFECT OF SELECTION ON DNA POLYMORPHISM

Natural selection affects the amount of DNA polymorphism, and the effect of selection on the estimate of the amount of DNA polymorphism depends on how the amount of DNA polymorphism is measured. Deleterious mutants are maintained in a population at low frequency. Since the number of segregating sites ignores the frequency of mutants, this number is strongly affected by the existence of deleterious mutants. On the other hand, the existence of deleterious mutants with low frequency does not affect the average number of nucleotide differences very much, since the frequency of mutants is considered. In the case of balancing selection such as overdominant selection, mutants can be maintained in a population at intermediate frequency, so that the effect of the selection on the average number of nucleotide differences might be larger than that on the number of segregating sites.

Kaplan *et al.* (1989) examined the hitchhiking effect at the DNA level, and concluded that in the region of low crossing-over the fixation of a selectively advantageous mutant at one nucleotide site can substantially reduce the amount of DNA polymorphism at linked sites. This study indicates that the amount of DNA polymorphism is determined by selection intensity and recombination rate as well as by mutation rate and effective population size.

Incidentally, Tajima (1990a) has shown that the expectation of the amount of DNA polymorphism is less when a mutation at a particular nucleotide site reaches fixation than at a random time, even if all newly arisen mutations are selectively neutral.

CONCLUSIONS

In this paper two measures of DNA polymorphism, *i.e.*, the number of segregating sites and the average number of nucleotide differences, were examined. When we analyze actual data, the number of segregating sites is recommended if the following conditions are satisfied: (1) genetic variation is caused only by neutral mutations, (2) the population is a random mating population, and (3) the population is at equilibrium. This is because the coefficient of variation of this number is smaller than that of the average number of nucleotide differences. On the other hand, if the above conditions are not satisfied, the average number of nucleotide differences is recommended because the number of segregating sites depends on sample size and this number cannot be transformed into a measure that is independent of sample size.

A statistical method for testing the neutral mutation hypothesis, developed by Tajima (1989b), was also summarized here. In this method it is assumed that the population from which a sample of DNA sequences is taken is a random mating and equilibrium population. Therefore, a significant deviation of D from 0 does not necessarily mean that the neutral theory is wrong. Before reaching a conclusion, we must consider whether the assumptions are correct. Table IV shows the expected value of D, $E(D)$, together with the relationship between the expectation of the average number of nucleotide differences, $E(k)$, and the expectation of the estimate of M obtained from the number of segregating sites, $E(\hat{M})$, under several conditions. When D is significantly

TABLE IV

Relationship between the Expectation of the Average Number of Nucleotide Differences, $E(k)$, and the Expectation of the Estimate of M Obtained from the Number of Segregating Sites, $E(\hat{M})$, and the Expectation of D, $E(D)$, under Several Conditions

Relationship between $E(k)$ and $E(\hat{M})$		$E(D)$
I. Random mating and equilibrium population		
(1) Neutral	$E(k) = E(\hat{M}) = 4Nv$	≈ 0
(2) Selection		
(i) Balancing selection	$E(k) > E(\hat{M}) > 4Nv$	> 0
(ii) Purifying selection	$E(k) < E(\hat{M}) < 4Nv$	< 0
II. Change in population size		
(1) Increase (N_0 to N_1)	$4N_0 v < E(k) < E(\hat{M}) < 4N_1 v$	< 0
(2) Decrease (N_0 to N_1)	$4N_0 v > E(k) > E(\hat{M}) > 4N_1 v$	> 0
III. Population subdivision		
No simple relationship between $E(k)$ and $E(\hat{M})$		

smaller than 0, there are the following possibilities: (1) purifying selection, (2) recent increase in population size (also bottleneck), and (3) population subdivision. On the other hand, when D is significantly larger than 0, the possibilities are: (1) balancing selection, (2) recent decrease in population size, and (3) population subdivision. If we study two or more genes or regions, however, we can often know whether selection is operating or not. That is, if change in population size or population subdivision is the cause of the significant deviation of D from 0, then we expect that the same deviation occurs for all genes or regions.

As has been shown in this paper, the amount of DNA polymorphism can be determined by a large number of factors and some of these factors were examined in this paper. It should be noted that we cannot identify the mechanism maintaining DNA polymorphism from the estimated amount of DNA polymorphism unless additional information is available. For example, the amount of DNA polymorphism smaller than the neutral expectation can be obtained not only when mutations are deleterious but also when there has been a bottleneck. One possible way to identify the mechanism is to use the distribution of nucleotide frequency in the sample. The expectation of this distribution, however, is known only in the case of neutral mutation. In this case the expected number of nucleotides whose frequency is i/n in a sample of n DNA sequences is given by

$$G_n(i) = M\left(\frac{1}{i} + \frac{1}{n-i}\right), \tag{31}$$

when $1 \le i \le n-1$ (Tajima, 1989b). We can make the substitution $M = S/a_1$ in Eq. (31) and compare the observed distribution of nucleotide frequency with the expected one, although we cannot conduct a significance test. In order to conduct a significance test, we must obtain the exact solution. Furthermore, it is necessary to know the distribution in cases other than the neutral mutation.

SUMMARY

A large amount of genetic variation is maintained in natural populations. There are many factors that determine the amount of genetic variation, such as mutation, selection, random genetic drift, migration, and others. In order to understand the mechanism of the maintenance of genetic variation, we must first estimate the amount of genetic variation.

There are two different quantities for measuring genetic variation at the DNA level, namely the number (S) of segregating (or polymorphic) nucleotide sites and the average number (k) of pairwise nucleotide differences among a sample of DNA sequences. These two measures have different statistical properties. The former depends on the sample size, so that we cannot use this measure directly. In the case where mutants are selectively neutral, $4Nv$ can be estimated by $S/\sum_{i=1}^{n-1}(1/i)$, of which the expectation is independent of the sample size, where N is the effective population size, v is the mutation rate per DNA sequence per generation, and n is the sample size. It should be noted that this measure is biologically meaningful *only* when mutants are selectively neutral. On the other hand, the latter measure gives the estimate of $4Nv$ in the neutral mutation model and has a clear biological meaning even if mutants are not neutral, although it has larger variance than does the value of $4Nv$ estimated from S.

The expected amount of DNA polymorphism has been examined under several models, including selection, migration, and change in population size, and the results indicate that these factors affect the amount of DNA polymorphism as expected. Unfortunately, we cannot identify the mechanism that maintains DNA polymorphism based only on the amount of DNA polymorphism estimated from a sample of DNA sequences.

Acknowledgment
I thank Dr. M. Nei for his valuable suggestions and comments on an earlier version of this paper.

REFERENCES

Hudson, R.R., M. Kreitman, and M. Aguadé, 1987. *Genetics* **116**: 153–159.
Kaplan, N.L., R.R. Hudson, and C.H. Langley, 1989. *Genetics* **123**: 887–899.
Kimura, M., 1968. *Nature* **217**: 624–626.
Kimura, M., 1983. *The Neutral Theory of Molecular Evolution.* Cambridge Univ. Press, Cambridge.
Lewontin, R.C., 1974. *The Genetic Basis of Evolutionary Change.* Columbia Univ. Press, New York.
Li, W.-H., 1976. *Theor. Popul. Biol.* **10**: 303–308.
Li, W.-H., 1977. *Genetics* **85**: 331–337.
Nei, M., 1975. *Molecular Population Genetics and Evolution.* North-Holland, Amsterdam.
Nei, M., 1987. *Molecular Evolutionary Genetics.* Columbia Univ. Press, New York.
Nei, M. and J.C. Miller, 1990. *Genetics* **125**: 873–879.
Slatkin, M., 1987. *Theor. Popul. Biol.* **32**: 42–49.
Strobeck, C., 1987. *Genetics* **117**: 149–153.
Tajima, F., 1983. *Genetics* **105**: 437–460.
Tajima, F., 1989a. *Genetics* **123**: 229–240.
Tajima, F., 1989b. *Genetics* **123**: 585–595.
Tajima, F., 1989c. *Genetics* **123**: 597–601.
Tajima, F., 1990a. *Genetics* **125**: 447–454.
Tajima, F., 1990b. *Genetics* **126**: 231–234.
Watterson, G.A., 1975. *Theor. Popul. Biol.* **7**: 256–276.

Estimating Selection Coefficients from the Phylogenetic History

BRIAN GOLDING

Department of Biology, York University, Toronto, Ontario, Canada M3J 1P3

Natural selection was proposed by Darwin (1859) to be the driving force behind evolution. Since that time, there has been much debate concerning the role of natural selection in shaping evolutionary change, particularly at the molecular level. The neutral theory of molecular evolution (Kimura 1968; see Kimura, 1983 for a review) has had a major influence on our concepts of evolution. This theory suggests that most molecular changes are more strongly influenced by random drift and by mutation than by selection. This forces us to be more precise and more quantitative when we invoke natural selection acting on a system.

Our ability to obtain useful estimates of selection coefficients has traditionally been rather poor (Manly, 1985). In part, this has been due to the excessive sample sizes required to achieve statistical significance and, in part, it is due to the poor quality of data that could be obtained. For example, electrophoresis can only identify electromorphs that may or may not correspond to individual alleles. The actual size of the selection coefficients may also be very small, such as the $s = 10^{-7}$ estimated by Sawyer *et al.* (1987) to be acting on *Escherichia coli*'s 6-phosphogluconate dehydrogenase.

The advent of new molecular techniques, including polymerase chain reaction (PCR) and DNA sequencing, promises a quantity and quality of data that may permit useful estimates of selection to be obtained. One way to estimate selection coefficients from these data is to examine recent transitions in sequences. A deterministic model that examined the most recent distinguishable ancestor of each extant sequence was developed by Golding *et al.* (1986). But this model did not include the confounding effects of random drift (see Iizuka, 1989) and requires inordinately large sample sizes to be statistically reliable (Golding, 1987).

A phylogeny contains a great deal of potentially useful information about the evolution of sequences. Besides the topological branching order, a phylogeny also contains information about the allele transitions, about the allele frequencies and about the inter-relationships between alleles (*e.g.*, whether a sequence variant covaries with other sequence variants). To make use of this store of information and to bypass the above problems it is desirable to have a maximum likelihood algorithm that would permit the reconstruction of phylogenies in the presence of variable selection coefficients as well as mutation and random drift. A completely general solution to this problem does not appear to be possible. However, several approximations can be made to permit exploration of the properties of such an algorithm (Golding and Felsenstein, 1990).

It is necessary to restrict the data to completely distinct species. This is necessary since the dynamics of two sequences that originate from within a single species are not independent of each other. Any selection may change the prior likelihood of phylogenies and this would greatly complicate the calculations. A second major problem is that the probabilities of an allelic transition are not easily calculated in the joint presence of mutation, drift, and selection. To work around this problem, two different approximations were developed in Golding and Felsenstein (1990). One approximation assumes that selection is a weak force and expresses a solution for the transition probabilities as a truncated power series in (Ns). The other approximation assumes that mutation is a much weaker force than selection and derives the transitions using the probabilities of eventual fixation.

In the following sections the derivation of these approximations will be illustrated and some properties of the algorithm will be explored through hypothetical examples.

APPROACH

The methodology is that of Golding and Felsenstein (1990). It is reviewed here in a shortened format to illustrate the essential features.

Consider sequences that are composed of sites with only two possible allelic states. These two states can be termed state "O" and state "X" for simplicity but they may correspond to any feature that can be encoded in a binary fashion. This would include the presence/absence of restriction sites or the presence/absence of any other feature or structure. Assume that the characters spontaneously mutate at a rate μ per gamete per generation and that the mutation rate to and from each state is equal. Generations are assumed to be discrete and the evolution of each site is assumed to be independent of all other sites.

If these allelic states are selectively neutral then their equilibrium frequencies are $1/2$ and the probabilities of a transition from one state to the other can be easily calculated. The probability that a site initially in state i will change to state j within time t is P_{ij}^t. The value of P_{ij}^t is given by

$$P_{ij}^t = 1/2 + 1/2e^{-2\mu t} \quad \text{if} \quad i = j$$
$$= 1/2 - 1/2e^{-2\mu t} \quad \text{if} \quad i \neq j.$$

It is assumed that mutations occur according to a Poisson process.

The likelihood of a tree can be found following Felsenstein's (1981) method. The likelihood is calculated in a recursive fashion working backward from the tips of the tree. The likelihoods of allelic states for nodes proximal to the tips are calculated first and these are then used to find the likelihood of states for more distal nodes. The likelihood of part of an evolutionary tree subtended by the k-th node for a site in state i is designated as $L_i^{(k)}$. Assume the tree to be strictly bifurcating with nodes m and n having a common ancestral node k. The time between node m and node k is t and between nodes n and k is t'. With these definitions the likelihood $L_i^{(k)}$ is

$$L_i^{(k)} = \left(\sum_{j=1}^{2} P_{ij}^t L_j^{(m)} \right) \left(\sum_{k=1}^{2} P_{ik}^{t'} L_k^{(n)} \right)$$

(Felsenstein, 1981). If nodes m and n designate extant species then the likelihoods $L_j^{(m)}$ and $L_k^{(n)}$ are known explicitly. The likelihoods will be either 1 or 0 depending on whether the extant species does or does not have that state (allele) at that particular site.

This process is repeated moving down the tree until the root is reached. Each successive step uses the likelihoods just calculated such that the value determined for $L_i^{(k)}$ is used to find the likelihood of the next node. The likelihood for the complete tree is found by summing the products of the root likelihoods with the prior probabilities of each state of the root. These prior probabilities of each state are usually taken to be their equilibrium frequencies.

This method can be modified to include the influences of selection on the sequence. This requires values for P_{ij}^t under the joint action of mutation, drift, and selection. This cannot be done in complete generality and instead was approached by Golding and Felsenstein (1990) using two approximations. The first approximation assumes that selection is weak relative to mutation and the second assumes that mutation is weak relative to selection.

For both approximations it is required that the individual taxa evolve independently after speciation. When selection occurs this is true only for trees consisting of different species. It is not true of individual, selected haplotypes within a species because selection may alter the prior probabilities of some phylogenies. Thus it is necessary that the effective population size be much less than the divergence time $(N_e \ll t)$ so that lineages coalesce rapidly when they are traced back within an ancestor. If polymorphism exists at the time of speciation, daughter species are formed from a randomly sampled haplotype. Note that the descendants must be contemporary in this model (assuming that selection has had equal opportunity to act in each taxon) and therefore a molecular clock is assumed.

The selection model considers genic selection with character "X" being potentially deleterious and the fitness of individuals carrying this gene reduced by a proportion 1-s. Each allelic state has the same selection coefficient independent of its site within the sequence and fitnesses are multiplicative.

A formula for the expected allele frequency when selection is acting in a finite population was first illustrated by Kimura (1955) and again by Avery (1978). Consider a population that has only small changes in allele "O" 's frequency, x, from one generation to the next. If this small change is designated as δx, then the expected n-th power of the allele frequency in the next generation is

$$E[(x+\delta x)^n] = E[x^n] + nE[x^{n-1}E_g(\delta x|x)] + n(n-1)/2E[x^{n-2}E_g(\delta x^2|x)] + \cdots,$$

The expectations E_g are over conceptually replicate populations each with a given value of x. For the above model,

$$E_g(\delta x|x) \simeq \mu(1-2x) + sx(1-x)$$
$$E_g(\delta x^2|x) \simeq x(1-x)/2N.$$

Substituting these and ignoring higher order terms gives

$$E(x+\delta x) \simeq E(x) + \mu(1-2E(x)) + sE(x) - sE(x^2)$$
$$E((x+\delta x)^2) \simeq E(x^2) + 2\mu(E(x) - 2E(x^2)) + E(x)/2N - E(x^2)/2N$$
$$+ 2sE(x^2) - 2sE(x^3).$$

This leads to an infinite series of equations unless $s = 0$. The first approximation used by Golding and Felsenstein (1990) solves these expectations as a power series in (Ns). For example, consider $E(x^n)$ expanded into

$$E(x^n) = a_n + b_n(Ns) + c_n(Ns)^2 + d_n(Ns)^3 + \cdots$$

where $a_n = E(x^n)|_{Ns=0}$ and the other terms must be solved. Substituting these leads to

$$
\begin{aligned}
E(x + \delta x) &\simeq E(x) + \mu(1 - 2E(x)) + s(a_1 + b_1(Ns) + \cdots) - s(a_2 + b_2(Ns) + \cdots) \\
&\simeq E(x) + \mu(1 - 2E(x)) + sa_1 - sa_2 \\
E((x + \delta x)^2) &\simeq E(x^2) + 2\mu(E(x) - 2E(x^2)) + E(x)/2N - E(x^2)/2N \\
&\quad + 2s(a_2 + b_2(Ns) + \cdots) - 2s(a_3 + b_3(Ns) + \cdots) \\
&\simeq E(x^2) + 2\mu(E(x) - 2E(x^2)) + E(x)/2N - E(x^2)/2N \\
&\quad + 2sa_2 - 2sa_3.
\end{aligned}
$$

Recursion equations are known for a_n since these are just the same equations with $s = 0$. Hence there is now a finite-dimensional system of equations (a 3-dimensional system to determine $E(x)$ or a 5-dimensional system to determine $E(x^2)$) that can be solved using standard techniques. However, this system does ignore terms of the order of $O(Ns)^2$. The system can be simply extended to solve for higher allele frequency moments and/or it can be extended to solve for more accurate solutions.

This system gives the expected change in allele frequency in one generation. The expression for t generations, $E^t(x)$, is found by iterating these equations. The transition probabilities are $P_{ox}^t = 1 - E^t(x)$, $P_{\infty}^t = E^t(x)$ with $E^o(x) = 1$, $E^o(x)|_{Ns=0} = 1$ and $E^o(x^2)|_{Ns=0} = 1$. Similarly $P_{xo}^t = E^t(x)$ and $P_{xx}^t = 1 - E^t(x)$ with initial conditions $E^o(x) = 0$, $E^o(x)|_{Ns=0} = 0$ and $E^o(x^2)|_{Ns=0} = 0$.

This is a linear, first order approximation and is only valid when Ns is small. For large Ns another approximation was used. This approximation assumes that mutation is weak relative to selection. The previous requirement that $N_e \ll t$ ensures that the time required for fixation of alternate alleles is very short relative to the length of time between fixations. Advantage can be taken of this condition by letting the fixation time approach zero relative to the time between fixations. The probability of eventual fixation for a particular advantageous allele is

$$U(s,Ns) = (1 - e^{-2s})/(1 - e^{-4Ns})$$

(Kimura, 1962). The probability of eventual fixation for a deleterious allele is $U(-s, -Ns)$. There are $2N\mu$ mutations of these alleles each generation. Using the approximation, the probability of a change in the allelic state per unit time is $a = 2N\mu U(s,Ns)$ or $b = 2N\mu U(-s, -Ns)$ depending on the direction of change.

Therefore the probability of a change from "O" to "X" in t generations is

$$P_{ox}^t = b/(a+b) \; (1 - e^{-(a+b)t})$$

and

$$P_{xo}^t = a/(a+b) \; (1 - e^{-(a+b)t})$$

while $P_{xx}^t = 1 - P_{xo}^t$ and $P_{oo}^t = 1 - P_{ox}^t$.

The two approximations to the transition probabilities are quite different. The series approximation is only valid with weak selection and includes the chance of polymorphism. This approximation is used when $4Ns < 0.1$. The weak mutation approximation considers the relative proportion of species fixed for alternate alleles and since mutation is weak, does not permit extended polymorphism. This approximation is more appropriate when selection is strong and is used when $4Ns > 0.1$. Both approximations require that $N_e \ll t$.

It was shown in Golding and Felsenstein (1990) that individual sites in a sequence will evolve independently and that the expectations for the product of allele frequencies will closely approximate the product of the expectations. There is no linkage disequilibrium generated between sites by this type of selection (Felsenstein, 1965; Birky and Walsh, 1988). Thus, the transition probabilities can be calculated treating each site independently (as if they were in linkage equilibrium) and the overall likelihood is the product of likelihoods from each site.

While the algorithm is conceptually capable of finding a maximum likelihood topology for a reconstructed phylogeny, in practice this requires excessive computation. This is not a great limitation of the algorithm since the null hypothesis in any test for selection is that $Ns = 0$. For this situation there are many excellent algorithms available to reconstruct phylogenies (reviewed in Felsenstein, 1988). Therefore, the tree topology will be assumed to be given for this algorithm.

EXAMPLES

1. Different Inferences from the Same Tree

This algorithm was applied to the hypothetical tree given in Fig. 1. This tree is based on a coalescent process with $2N_e = 10^4$ and a sample of 30 species (using the algorithm of Hudson, 1983). Two of the 30 species were arbitrarily chosen to contain potentially selected characters (X) while the remaining 28 species contain selectively neutral characters (O). The two species chosen are listed in Fig. 1 in line A. Since only 2 of 30 species contain character "X" and

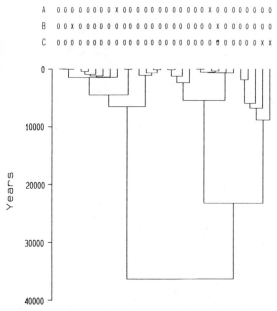

A 0 0 0 0 0 0 0 0 X 0 0 0 0 0 0 0 0 0 0 0 X 0 0 0 0 0 0 0 0

B 0 0 X 0 0 0 0 0 0 0 0 0 0 0 0 0 0 0 0 0 0 X 0 0 0 0 0 0 0 0

C 0 0 0 0 0 0 0 0 0 0 0 0 0 0 0 0 0 0 0 0 0 0 0 0 0 0 0 X X

Fig. 1.   A tree based on a coalescent for 30 genes with $2N = 10^4$. Three alternate distributions of the alleles "O" and "X" are shown at the top and identified by lines A, B, and C.

since the expected equilibrium frequency of "X" is $1/2$, this would suggest that something has constrained the evolution of this character. This could be either a lack of mutation, very close ancestral relationships among the species or selection. The mutation rate for the characters in Fig. 1 is set at $\mu = 0.5 \times 10^{-5}$ so that $4N\mu = 0.1$ and is not limiting. The branch lengths are those determined from the coalescent and are lengths that are appropriate for sequences originating from different individuals of one species rather than from different species. Hence this should represent a worst case in the sense that the species are more closely related than normally possible.

The likelihood of this tree with different values of $4Ns$ is shown in Fig. 2 as curve $(\times 10^0)$. The likelihood of the tree decreases as selection increases against the "X" character. Hence there is no support for selection acting against this character. If, however, all branch lengths are multiplied by 10 (curve $\times 10^1$) then the likelihood of the tree increases even in the absence of selection (when $4Ns = 10^{-3}$). This increase is due to the higher likelihood of mutations occurring if the branch lengths are longer. As selection increases the likelihood of the tree increases and reaches a maximum when $4Ns \approx 2$. This

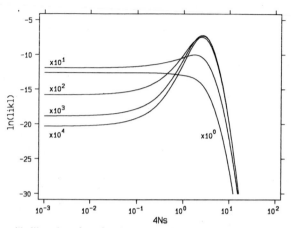

Fig. 2. The $\log_e$ likelihood surface for the tree in Fig. 1, with pattern A. Branch lengths of the tree in Fig. 1 have been increased 10^0 to 10^4 fold.

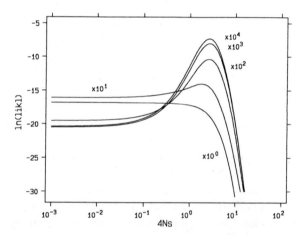

Fig. 3. The $\log_e$ likelihood surface for the tree in Fig. 1, with pattern B. Branch lengths of the tree in Fig. 1 have been increased 10^0 to 10^4 fold.

increase is due to the increased likelihood of observing only 2 of 30 species with a deleterious character. The likelihood eventually falls off again as the selection increases further since the probability of obtaining even 2 of 30 becomes small if the character is too deleterious.

As the branch lengths increase further (curves $\times 10^2$, $\times 10^3$, $\times 10^4$) and in the absence of selection, the likelihood of the tree decreases. With longer branch lengths the number of species with an "X" character should be closer

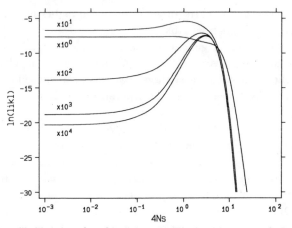

Fig. 4.   The $\log_e$ likelihood surface for the tree in Fig. 1, with pattern C. Branch lengths of the tree in Fig. 1 have been increased 10^0 to 10^4 fold.

to its equilibrium expectation. Again as the $4Ns$ increases, the likelihood of the tree increases with a maximum at $4Ns \approx 2.5$. Note that the maximum likelihood is similar whether the branch lengths are increased 10^2 or 10^4 fold.

Another pair can be randomly chosen as in Fig. 1, line B. In this case the likelihood of the tree with different branch lengths and different values of $4Ns$ is shown in Fig. 3. Here, the likelihoods are generally smaller since one of the species with character "X" is very closely related to a species with character "O". Such a rapid transition is unlikely and reduces the overall level of the likelihoods. As the branch lengths are increased the likelihoods in Fig. 2 and Fig. 3 become more similar. This time however, the maximum likelihood increases slowly as the branch lengths are increased 10^2 to 10^4 fold.

A non-random pair of species can also be chosen to contain the "X" characters. If the species with the two largest branches (the two species at the extreme right in Fig. 1) are chosen, the likelihoods are as shown in Fig. 4. This figure provides quite a different response to selection. In the absence of selection, the likelihoods are much larger (since the two allelic transitions now occur on long branches). Again, in the absence of selection ($4Ns = 10^{-3}$), the likelihood decreases as the branch lengths increase. This time, however, the curve with the largest maximum occurs when the branch lengths are increased only 10 fold. This was unexpected and is contrary to the situations in Figs. 2 and 3 since it shows that a higher overall likelihood can be obtained with shorter branch lengths and that this likelihood is larger than that in Figs. 2 and 3 for any branch lengths. Again contrary to the situations in Figs. 2 and 3, when selection is very strong, the original tree has a higher likelihood than the trees with increased branch lengths.

These figures were all generated using a single tree, all have the same mutation rate and all have two "X" characters. Yet small changes in the placement of the species that carry these characters can cause large changes to the behavior of a likelihood surface incorporating selection.

2. The MLE of 4Ns for Distantly Related Species

At the other extreme from having closely related species would be a situation with species that are distantly related. This case permits a great simplification since the phylogeny no longer has to be included in the analysis and the likelihood will simply be a product of the expected equilibrium allele frequencies. One question that is of interest for such data is how the maximum likelihood estimate (MLE) and its significance changes as the frequency of the allele changes. This is shown in Table I for a sample of 50 unrelated species. Table I gives the MLE of 4Ns and shows that it does not decrease linearly with a linear increase in the frequency of the deleterious character. Instead, the maximum is achieved with very low frequencies and the MLE of 4Ns quickly decreases with increasing frequency and then slowly levels off as the frequency increases further. Also included in this table is the likelihood of the observation in the absence of selection. Note that even though the estimate of 4Ns may be quite small (as with, say, 10/50 where the MLE of $4Ns = 1.3864$) this level of selection can still be very significant.

One of the advantages of a likelihood approach is that statistical evalua-

TABLE I

Likelihood for Distantly Related Species with Sample Size Held Constant.

Frequency of deleterious character	MLE of 4Ns	ln(Likl)	ln(Likl) at $4Ns=0$	$-2\ln(LR)$	Prob.
1/50	3.8922	−4.9020	−34.6574	59.511	$<10^{-9}$
2/50	3.1784	−8.3972	−34.6574	52.520	$<10^{-9}$
3/50	2.7518	−11.3484	−34.6574	46.618	$<10^{-9}$
4/50	2.4426	−13.9385	−34.6574	41.438	$<10^{-9}$
5/50	2.1974	−16.2541	−34.6574	36.807	1.80×10^{-9}
6/50	1.9926	−18.3462	−34.6574	32.622	1.24×10^{-8}
7/50	1.8155	−20.2482	−34.6574	28.818	7.95×10^{-8}
8/50	1.6584	−21.9835	−34.6574	25.348	4.79×10^{-7}
9/50	1.5165	−23.5697	−34.6574	22.175	2.49×10^{-6}
10/50	1.3864	−25.0201	−34.6574	19.275	1.13×10^{-5}

TABLE II
Likelihoods for Distantly Related Species with Frequency Held Constant.

Frequency of deleterious character	MLE of $4Ns$	ln(Likl)	ln(Likl) at $4Ns=0$	$-2\ln(LR)$	Prob.
2/10	1.3864	-5.0040	-6.9315	3.855	4.96×10^{-2}
3/15	1.3864	-7.5060	-10.3972	5.782	1.62×10^{-2}
4/20	1.3864	-10.0080	-13.8629	7.710	5.49×10^{-3}
5/25	1.3864	-12.5101	-17.3287	9.637	1.91×10^{-3}
6/30	1.3864	-15.0121	-20.7944	11.565	6.72×10^{-4}
7/35	1.3864	-17.5141	-24.2602	13.492	2.40×10^{-4}
8/40	1.3864	-20.0161	-27.7259	15.420	8.61×10^{-5}
9/45	1.3864	-22.5181	-31.1916	17.347	3.11×10^{-5}
10/50	1.3864	-25.0201	-34.6574	19.275	1.13×10^{-5}

TABLE III
Estimates of $4Ns$ with Increasing Sample size.

Frequency of deleterious character	MLE of $4Ns$	ln(Likl)
1/10	2.1974	-3.2508
1/15	2.6393	-3.6740
1/20	2.9447	-3.9703
1/25	3.1784	-4.1986
1/50	3.8922	-4.9020
1/100	4.5955	-5.6002
1/1,000	6.9074	-7.9073
1/10,000	9.2112	-10.2103

tion of the strength of selection can be made using standard likelihood ratio tests. Twice the negative logarithm of the likelihood ratio is asymptotically distributed as a Chi-Square with one degree of freedom. The probability that the observed frequencies are due to chance is shown in the last column in Table I.

The change in the significance in the estimate of $4Ns$ as a function of sample size is shown in Table II. Here, different overall sample sizes are compared when the frequency of the deleterious characters is kept constant at 20%. Because the frequency is kept constant and there is no complicating phylogeny considered here, the MLE of $4Ns$ is a constant 1.3864. But the significance of this observation changes dramatically. With a sample of 10 species the probability that this biased observation is due to chance is 5%, barely significant. With a sample size of 50 however, the same probability is 0.001%.

It is also interesting to examine a situation where the number of species carrying a deleterious character is kept constant at 1 but with the sample size increasing. This is shown in Table III. With 1 of 10 the maximum likelihood estimate of $4Ns$ is 2.2 and is just significantly different from zero. As the sample size increases the maximum likelihood estimate of $4Ns$ also increases but only extremely slowly. Indeed, even a frequency of 1 in 10,000 for the deleterious character yields an estimate of $4Ns$ that is still less than 10. This demonstrates the strength and effectiveness of natural selection.

3. Unusual Patterns in Phylogenies

In addition to rare characters being potentially under selection, some unusual patterns within a phylogeny may suggest selection is acting. One such pattern is found when the characters are equally represented but are present in unrelated species. A simple bifurcating tree with such a pattern is shown in Fig. 5. This is an idealized pattern that one might expect with overdominance. There is a high frequency of both characters but their pattern of change in the phylogeny suggests either multiple origins or their joint presence in all ancestral species.

The likelihood of the tree in Fig. 5 is shown in Fig. 6 for three different mutation rates. The length of branches between nodes in Fig. 5 is set to $t = 10^5$. The likelihood surfaces in Fig. 6 assume $2N = 10^4$, $4N\mu = 0.01$ (solid line), $4N\mu = 0.001$ (dotted line) and $4N\mu = 0.0001$ (dash-dot line). In all cases the likelihood is increased by selection. The increase in the likelihood is larger than in Figs. 2–4 (note the scale change). This increase in the likelihood only occurs because the mutation rate is too small to accommodate the number of changes

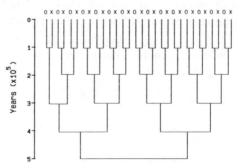

Fig. 5.   A bifurcating tree with 30 species and a dispersed pattern of alleles "O" and "X" with frequencies 1/2.

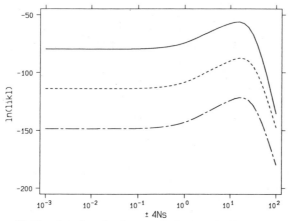

Fig. 6.   The $\log_e$ likelihood surface for the tree in Fig. 5 with $4N\mu = 0.01$ (solid line), 0.001 (dashed line), and 0.0001 (dash-dot line).

implied by the phylogeny. With larger mutation rates than those shown here, the overall level of the likelihood is larger and there is a monotonic decline in the likelihood as selection is introduced. The dependence of the likelihood on mutation rate is also shown by the variable position of the maximum in the curves. The maximum of the likelihood surfaces is shifted to the right in Fig. 6 with decreasing mutation rates.

The distribution of deleterious characters in Fig. 6 is symmetrical and the two allelic states can be interchanged without affecting the nature of the phylogeny. Thus, the likelihood surfaces with s either positive or negative are equal. The phylogeny in Fig. 5 supports the influence of selection both with "X" as a deleterious character and with "X" as an advantageous character. Both selection schemes provide a statistically better "fit" than does strict neutrality.

The likelihoods of the ancestral states are also obtained by this algorithm. Perhaps contrary to expectation, it is the deleterious character that has a higher likelihood as the ancestral state. The character "X" is the most likely ancestral state despite the fact that it may be very much more deleterious than the alternate allelic state. The reason for this is that the transition from a deleterious state to a neutral (or relatively advantageous) allelic state is much more likely than the transition from a neutral state to a deleterious state. Thus, if a character is maintained in high frequency by a mutation-selection balance and the sample for some species includes this deleterious character, then it is this state that a maximum likelihood approach will favor as the ancestral state.

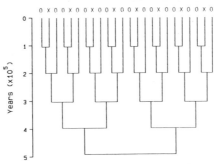

Fig. 7. A tree with 24 species and a dispersed pattern of alleles "O" and "X" with frequencies 2/3 and 1/3, respectively

A tree can be chosen that forces the ancestral state to be the neutral character. A simple tree that implies this is given in Fig. 7. Here, there is again a high frequency of both allelic states but the pattern of the phylogeny suggests that the neutral character is ancestral. This pattern is no longer symmetrical and the shape of the likelihood surfaces depend on whether s is positive or negative. The likelihood surface with s positive is shown in Fig. 8a and with s negative in Fig. 8b for $2N = 10^4$, $4N\mu = 0.01$, 0.001, and 0.0001.

Selection acting with "X" as a deleterious character is not supported (Fig. 8a) but selection acting with "X" as an advantageous character is supported (Fig. 8b). Again, in both these situations, if selection is strong enough, the deleterious character is favored for the ancestral state. This causes the unusual shape to the likelihood surface in Fig. 8a as the ancestral state changes from "O" (with $4Ns$ small) to "X" (with $4Ns$ large).

CONCLUSIONS

One of the major problems in estimating selection coefficients is a deficiency of large enough sample sizes to give statistical validation to the estimates. The advent of techniques that permit rapid sequence determination should permit much larger sample sizes. In this case, while the actual number of individuals sampled may decrease, the number of sites potentially compared can increase by orders of magnitude. Due to the independence of these sites, data can be combined for each site and the realized sample size can be easily increased several orders of magnitude. This should make it possible to detect even very small selection coefficients by considering longer sequences.

To attempt to extract as much information as possible from sequence

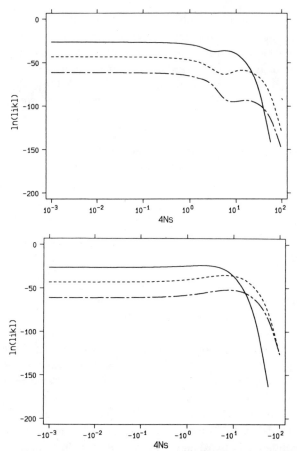

Fig. 8.   The $\log_e$ likelihood surface for the tree in Fig. 7 with $4N\mu = 0.01$ (solid line), 0.001 (dashed line), and 0.0001 (dash-dot line) and with $4Ns$ positive (8a) or negative (8b).

data, we have developed an algorithm that finds maximum likelihood estimates of selection coefficients. Some properties of this algorithm are examined here using three hypothetical examples. Each of the examples shows how unusual features can emerge and illustrate collectively that the interplay between selection, drift, mutation, and a phylogeny can be quite complicated.

The likelihood surfaces in Figs. 2 and 3 are quite different. Yet both have the same phylogeny and differ only by a random choice of 2 out of 30 species to contain the deleterious "X" character. In fact, since one of the "X" characters in Fig. 2 and Fig. 3 comes from closely related, sibling species, the two figures actually differ only by the choice of a single species. Thus, depending on the species with deleterious characters (or on the phylogeny) and depending on the branch lengths (or on the mutation rates), the presence of selection may or may not be supported even for superficially similar data.

While small selection coefficients can be estimated, the algorithm appears to be less efficient at estimating very large coefficients. Even with frequencies as low as 10^{-4}, the MLE of $4Ns$ for unrelated samples is still less than 10 (Table III). While the variance around the estimate is very large and somewhat skewed toward higher values of $4Ns$, nevertheless this strength of selection is sufficient to cause the biased allele frequency. Note that this algorithm does not consider polymorphisms and a method that permits multiple samples within a species might yield somewhat different results. Additionally, when more closely related species are considered the maximum likelihood estimates of $4Ns$ may be larger. Still, such a low estimate of $4Ns$ for such a strongly biased observation is unusual given the selection coefficients commonly invoked in other studies. For example, Takahata and Nei (1990) estimate that $Ns = 1,000$ or larger for overdominant selection acting on some major histocompatibility complex (MHC) polymorphisms. This magnitude of selection seems to be well outside the realms of possibility for this algorithm.

The example phylogenies given in Figs. 5 and 7 are the types of pattern that one might expect for overdominant selection. Again, the likelihood surfaces for these phylogenies are unusual. For the phylogeny in Fig. 5 both selection for and against the "X" character is supported but for the phylogeny in Fig. 7 only selection favoring the character "X" is supported.

It is also interesting that the most likely ancestral states are the deleterious states. This tendency becomes stronger as the selection becomes stronger and is due to the differences between the likelihoods of fixing a deleterious character and fixing an advantageous character. This has interesting implications for phylogenies based on character data and with ancestral states reconstructed using standard parsimony algorithms. In the presence of selection, the ancestral states will definitely *not* be those implied by a parsimony algorithm. Since most of the character states commonly considered by these studies are assumed to be selectively important, it questions the desirability of reconstructing trees without an underlying evolutionary model.

This algorithm can be extended in many ways. It is possible to permit mutation rates that are not symmetrical to/from the different allelic states and it is possible to derive a k-allele model that permits DNA/RNA sequences to be studied. These studies are currently in progress.

SUMMARY

Distinctive patterns of genetic change occur in phylogenies when selection acts on DNA sequences. Deleterious selection, in particular, strongly reduces the expected frequency of alleles and makes the transition from deleterious to neutral alleles less likely than the reverse transition. These patterns of change in a phylogeny can be analyzed and used to search for the influence of selection. The likelihood for different phylogenies is explored to determine the properties of a likelihood surface both with and without selection. This permits the application of a standard likelihood ratio test to search for selection. Because the phylogenetic history of a sample can extend over millions of years, the effects of selection have a long period of time to accumulate. Thus, the algorithm can detect even weak selection if it has been acting for a sufficient time period and given a sufficient sample size. While large samples are often required to detect selection, this requirement can be bypassed using the large number of sites available from a small sample of sequences. Unusual distributions of alleles between species, such as an alternating pattern of neutral/deleterious alleles, are also examined.

Acknowledgments

I thank Dan Fieldhouse for his comments on a draft of this manuscript. This work was supported by Natural Sciences and Engineering Research Council of Canada grant number U0336 and by the Canadian Institute for Advanced Research.

REFERENCES

Avery, P.J., 1978. *Theor. Popul., Biol.* **13**: 24–39.
Birky, G.W. and J.B. Walsh, 1988. *Proc. Natl. Acad. Sci. U.S.A.* **85**: 6414–6418.
Darwin, C., 1859. *On the Origin of Species.* Facsimile edition (1964), Harvard University Press, London.
Felsenstein, J., 1965. *Genetics* **52**: 349–363.
Felsenstein, J., 1981. *J. Mol. Evol.* **171**: 368–376.
Felsenstein, J., 1988. *Annu. Rev. Genet.* **22**: 521–565.
Golding, G.B., 1987. *Genet. Res. Camb.* **49**: 71–82.
Golding, G.B., C.F. Aquadro, and C.H. Langley, 1986. *Proc. Natl. Acad. Sci. U.S.A.* **83**: 427–431.
Golding, G.B. and J. Felsenstein, 1990. *J. Mol. Evol.* **311**: 511–523.
Hudson, R.R., 1983. *Evolution* **37**: 203–217.

Iizuka, M., 1989. *Genet. Res.* **54**: 231-237.

Kimura, M., 1955. *Proc. Natl. Acad. Sci. U.S.A.* **41**: 144-150.

Kimura, M., 1962. *Genetics* **4**: 713-719.

Kimura, M., 1968. *Nature* **217**: 624-626.

Kimura, M., 1983. *The Neutral Theory of Molecular Evolution.* Cambridge University Press, New York.

Manly, B.F.J., 1985. *The Statistics of Natural Selection.* Chapman and Hall, Ltd. New York.

Sawyer, S.A., D.E. Dykhuizen, and D.L. Hartl, 1987. *Proc. Natl. Acad. Sci. U.S.A.* **84**: 6225-6228.

Takahata, N. and M. Nei, 1990. *Genetics* **124**: 967-978.

Evolutionary Inferences from Molecular Characterization of Self-Incompatibility Alleles

ANDREW G. CLARK

Institute of Molecular Evolutionary Genetics, Department of Biology, Pennsylvania State University, University Park, PA 16802, U.S.A.

The term gametophytic self-incompatibility is ascribed to flowering plants that recognize the genotype of the gametophyte (pollen) and abort pollen tube growth if the allele borne by the pollen matches either allele of the pistil. It has long been recognized that the locus that determines the identity of both the pollen and the pistil (the S-locus) is transmitted as a single Mendelian gene (East and Mangelsdorf, 1925), although the magnitude of the rejection reaction is somewhat variable, and other loci modify the strength of the reaction to some degree (de Nettancourt, 1977). By diallel crosses, it was established in the 1930's that the number of distinct S-alleles could be very large, even in small populations. For example, a rare *Oenothera* species whose entire population was estimated at around 500 individuals contained 45 distinct S-alleles (Emerson, 1938, 1939; Lewis, 1948). These observations motivated the development of population genetic theory designed to address the expected number of alleles maintained in a finite population in the face of mutation, random genetic drift, and the selection mediated by self-incompatibility (Wright, 1939, 1960, 1964; Fisher, 1958; Ewens, 1964, 1969; Nagylaki, 1975; Yokoyama and Nei, 1979; Yokoyama and Hetherington, 1982). This theory was somewhat controversial, with the controversy centering around some of the approximations that were made in order to apply diffusion theory, but the approximations of Fisher and Wright have subsequently been shown to be very similar over the important range of the parameter space. The theory most relevant to the sequence data that are currently being collected makes predictions about the genealogy of S-alleles. We will develop some of this theory and point out how the gene genealogy of an S-locus is distinguished from that of an overdominant locus.

Population genetic theory designed to address the origins of self-incompatibility has meanwhile shed light on the aspects of the plant breeding system that promote or retard invasion of self- incompatibility into a self-compatible population. Because a gametophytic self-incompatibility locus requires a minimum of three alleles, it is clear that if self-incompatibility were to evolve from a self-compatible species, there must be a period of co-occurrence of both types of plant in the population. The question that theoretical population geneticists have asked is, under what conditions will a novel self-incompatibility allele be able to increase in frequency in such a population? Charlesworth and Charlesworth (1979) examined models that describe the changes in frequency of both active and inactive self-incompatibility alleles, where inactive alleles allow self-fertility. They found that the self-fertility allele can invade a self-incompatible population provided the number of S-alleles is below a threshold, and the threshold number of S-alleles depends on the rate of selfing. The conditions for invasion of self-compatibility depend on whether the self-compatibility arises from elimination of pollen function or of stigma function. Invasion of the self-incompatibility allele into a self-compatible population requires very strong inbreeding depression, such that outcrossed progeny are twice as fit as progeny that result from self-fertilization.

The two essential components of any model for the evolution of breeding systems are parent-offspring relatedness and offspring quality. In a series of papers, Uyenoyama (1988a, b, c, 1991) has developed the idea that self-incompatibility has evolved as a means of preferential maternal investment in offspring of high quality. Two features that are necessary for invasion of self-incompatibility include the existence of inbreeding depression and some outcrossing. This assures that the outbreeding generated by self-incompatibility will improve offspring fitness (Uyenoyama, 1988a). Early self-incompatibility systems may have resulted in only partial inhibition of pollen tube growth, and although a stronger reaction increases the outbreeding of the progeny, it also results in lower pollen success. Conditions for invasion of weak self-incompatibility depend on the strength of the rejection reaction, as well as the levels of inbreeding, inbreeding depression, recessivity of stylar expression, and associations between the S-locus and loci causing inbreeding depression (Uyenoyama, 1988b). The evolution of a complete self-incompatibility reaction from weak incompatibility was modeled by allowing a modifier locus to control the strength of the reaction (Uyenoyama, 1988c). The modifier allele that intensifies the self-incompatibility reaction invades the population provided the mean fitness of the offspring is increased. A three-locus model, with one proto-S-locus, one locus that modifies the strength of the incompatibility reaction, and one overdominant viability locus, revealed conditions under which self-incompatibility can invade even when inbreeding depression is considerably weaker than the 2-fold reduction in fitness that Charlesworth and Charlesworth (1979) found.

A deeper understanding of the evolution of the S-locus will ultimately require an elucidation of the cellular mechanism for the recognition of pollen and for the aborted pollen tube growth. Although these mechanisms are still unclear, the recent cloning of the S-gene in both sporophytic and gametophytic systems has laid the foundation for eventual elucidation of the cellular processes involved. Molecular evolutionary analysis of S-allele sequences has also demonstrated the unique evolutionary history of the S-locus. This chapter will focus on the gametophytic system, but it should be noted that S-alleles from plants that exhibit sporophytic self-incompatibility, such as *Brassica oleracea*, have also been cloned and sequenced (Nasrallah *et al.*, 1985, 1987). There appear to be multiple S-loci in the sporophytic system (Boyes *et al.*, 1991), and they too exhibit unusually high sequence diversity (Nasrallah and Nasrallah, 1989).

MOLECULAR CHARACTERIZATION OF GAMETOPHYTIC S-ALLELES

The genetic basis of gametophytic self-incompatibility has been most studied in the family Solanaceae. The pistil protein products of a number of S-alleles have been identified based on their cosegregation with respective alleles in genetic crosses. Cloning of the S-cDNA was facilitated by the great abundance of the S-proteins and their pistil-specific localization. A number of S-proteins of *Nicotiana alata* and *Lycopersicon peruvianum* were purified and their amino-terminal sequences determined. The first cDNA clone for an S-proteins was isolated from *N. alata* by Anderson *et al.* (1986). Degenerate oligonucleotides corresponding to the N-terminal sequence of the S_2-protein of *N. alata* were used to identify the S_2-cDNA from pistil-specific cDNA libraries. The cDNA clone for the S_z allele was not obtained by probing with the S_2 clone, but rather it was found to be the most abundant class of pistil specific cDNA clones (Kheyr-Pour *et al.*, 1990). The S_z clone was then used as a probe to identify additional S-allele cDNA clones (Xu *et al.*, 1990a). Subsequently, cDNA for more alleles of *N. alata* and other solanaceous species were isolated using heterologous probes (Ai *et al.*, 1990; Xu *et al.*, 1990b).

When DNA sequence databases were queried for matches with the
S-allele sequences, a remarkable homology was found to a fungal RNase
(McClure *et al.*, 1989). Subsequently, the S-proteins themselves were found to
exhibit ribonuclease activity (Singh *et al.*, 1991), and these observations
stimulated much speculation about how the rejection reaction might be
mediated by RNA degradation. McClure *et al.* (1990) tested whether pollen
RNA is differentially degraded in compatible *vs.* incompatible styles by labeling
pollen in plants grown in ^{32}P. The results indicate that pollen RNA is degraded
in incompatible styles, but not in compatible styles. Pollen does not synthesize
rRNA as the pollen tube grows, so degradation of rRNA would result in loss
of protein synthesis and arrestment of pollen tube growth (Haring *et al.*, 1990).
While these results do not address the nature of the allelic specificity for the
rejection reaction, nor do they disprove the possibility that the RNA degradation might be a byproduct of another inhibitory mechanism, the strong
conservation of sequence (11 consecutive amino acids are conserved in a block
that is shared with the fungal RNase) clearly indicates that this region encodes
a critically important function.

Genomic clones for two S-alleles each of *Petunia inflata* and *Solanum
tuberosum* (Ai *et al.*, 1991) and the S_2 alleles of *S. tuberosum* (Kaufmann *et al.*,
1991) have been reported. All contain a small intron (92 to 114 bp) located in
the hypervariable region HVa. A dotplot of the S_1 and S_3 alleles of *P. inflata*
reveals a remarkable loss of sequence homology 5′ from the gene (Coleman and
Kao, 1991). Within this virtually totally diverged region, at a position − 30 bp
5′ from the initiation of transcription is a conserved TATA box. The intron
retains a trace of homology, but in comparison to introns of other genes, they
too are extraordinarily divergent. Although this magnitude of sequence divergence is remarkable, the data from genomic Southern blots and genetic
segregation tests strongly suggest that *Petunia* has a single S-gene. Taken
together, these observations suggested to us that the region flanking the S-locus
is devoid of recombinational exchange (Clark and Kao, 1991), and that the
flanking region has accumulated divergence since the common ancestor of the
S_1 and S_3 alleles. Southern blot analysis of the flanking regions has shown that
they are highly repeated, but the sequences are not heterochromatic, and bear
no similarity to any known transposable elements (Coleman and Kao, 1991).
Further analysis of the genomic region flanking the S-locus is underway.

MOLECULAR COMPARISON OF S-ALLELES

To date, amino acid sequences of 21 S-proteins from six solanaceous species representing four genera have been reported, all of which were inferred from the corresponding cDNA sequences. Sequences of alleles from N. *alata* included S_{FII}, S_2, S_a, and S_{Inic} from Kheyr-Pour *et al.* (1990), and S_{2nic}, S_{3nic}, and S_{6nic} from Anderson *et al.* (1989). Alleles from *P. inflata* included S_{1pet}, S_{2pet}, and S_{3pet}, taken from Ai *et al.* (1990), and alleles of *Petunia hybrida*, including *PS1B*, *PS2A*, and *PS3A*, were obtained from Clark *et al.* (1990). Alleles S_o and S_x are from a strain of self-compatible *P. hybrida* (Ai *et al.*, 1991). *Solanum chacoense* alleles (taken from Xu *et al.*, 1990a, b) included S_{2sol} and S_{3sol}, and *S. tuberosum* alleles included *S1tub*, *S2Stub*, and *Sr1Stub* (Kaufmann *et al.*, 1991). The sequence of allele S_{5lyc} of *Lycopersicon esculentum* was obtained by Tsai *et al.* (submitted). The amino acid sequences of these 21 alleles excluding the leader peptide were aligned with the optimization algorithm of Lipman *et al.* (1989).

Comparison of these sequences has revealed primary structural features of the S-protein which might shed light on its function in recognition and rejection of self-pollen. The multiple sequence alignment revealed five conserved and two hypervariable regions (Fig. 1). The two hypervariable regions are the most hydrophilic segments of the S-protein, suggesting a possible role in the specific interaction with its counterpart S-allele product in pollen. Other

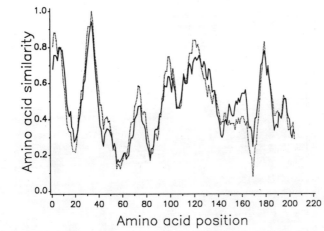

Fig. 1. Average pairwise amino acid similarity for the 21 S-alleles from six species of Solanaceae in a sweeping window of 10 amino acids (solid line) and for the comparison restricted to the Nicotiana alleles. Included in the analysis are seven N. *alata* alleles, three alleles from *P. inflata*, three alleles of *P. hybrida*, two alleles from a strain of self-compatible *P. hybrida*, two alleles from *S. chacoense*, and three alleles of *S. tuberosum*.

TABLE I

Matrix of Percent Pairwise Amino Acid Identity of S-alleles

	Sz	Sa	S1nic	S2nic	S3nic	S6nic	S5lyc	S1pet	S2pet	S3pet	PS3A	PS2A	PS1B	Sx	So	S2sol	S3sol	S1Stb	Sr1St	S2Stub
SF11	62.05	40.00	46.81	45.74	48.42	46.56	60.31	55.67	55.15	54.40	53.89	54.12	43.52	60.62	42.49	43.75	45.03	51.81	51.03	60.82
Sz		40.70	47.12	44.50	48.70	47.92	71.21	65.15	63.32	67.34	63.82	62.81	43.88	71.07	44.90	44.39	46.67	48.45	47.18	70.71
Sa			48.80	46.99	52.98	51.50	40.94	39.18	40.12	42.11	39.77	40.12	46.20	39.64	47.37	49.71	43.53	42.60	44.71	42.44
S1nic				67.71	69.27	62.50	45.03	43.23	41.88	43.46	41.58	41.36	46.35	45.50	46.88	47.64	52.36	43.39	43.68	44.74
S2nic					65.10	61.98	42.93	40.10	40.84	41.36	38.95	43.23	41.97	45.83	47.12	49.22	52.33	47.64	45.26	43.68
S3nic						71.50	47.67	45.36	44.04	44.56	43.23	41.36	46.63	47.94	50.00	44.74	45.13	48.45	47.40	48.96
S6nic							47.92	44.04	42.19	43.23	41.36	40.10	42.49	46.84	46.63	45.64	45.13	48.45	43.98	48.69
S5lyc								59.90	58.88	60.41	60.71	57.87	44.10	69.39	45.64	45.13	48.21	45.64	44.90	85.93
S1pet									74.75	74.87	73.10	73.47	65.82	65.99	40.31	45.64	44.85	45.13	48.21	60.71
S2pet										81.31	87.37	93.47	82.32	65.99	41.84	41.33	46.15	44.85	45.13	60.41
S3pet											79.29	82.32	65.99	65.13	42.05	41.54	44.10	44.62	44.62	60.71
PS3A												88.89	40.51	65.13	41.03	41.54	45.88	45.60	45.36	61.22
PS2A													41.33	63.27	39.29	44.62	43.81	43.59	44.62	58.88
PS1B														41.45	83.50	41.97	42.86	43.59	44.62	42.35
Sx															41.97	44.56	46.11	53.13	50.78	71.28
So																42.86	42.56	43.08	43.88	44.39
S2sol																	42.56	44.79	45.60	46.67
S3sol																		43.75	46.11	45.88
S1Stub																			88.44	52.06
Sr1Stub																				50.26

notable conserved residues are the eight cysteines, all of which are most likely involved in intramolecular disulfide bonds, as indicated by the finding of four disulfide bonds in the S-protein of *P. inflata* (Ai and Kao, unpublished results). These eight cysteines are also conserved in an extracellular ribonuclease RNase LE, identified in *L. esculentum* a self-compatible species (Jost *et al.*, 1991). Two

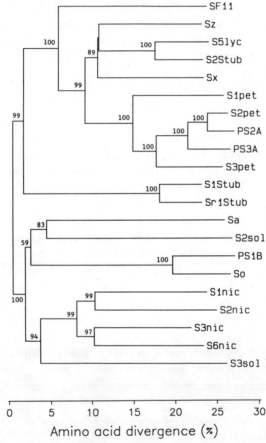

Fig. 2.   Neighbor-joining tree of the 21 *S*-alleles constructed from the pairwise distances between amino acid sequences (Table I). Numbers at each node indicate the percentage of 1,000 bootstrap samples whose tree contained the indicated node. *N. alata* alleles include S_{F11}, S_z, S_a, and S_{1nic}, S_{2nic}, S_{2nic}, and S_{2nic}. Alleles from *P. inflata* include S_{1pet}, S_{2pet}, and S_{3pet}, and alleles of *P. hybrida* include *PS1B*, *PS2A*, and *PS3A*. Alleles S_o and S_x are from a strain of self-compatible *P. hybrida*. *S. chacoense* alleles include S_{2sol} and S_{3sol}, and *S. tuberosum* alleles include *S1tub*, *S2Stub*, and *Sr1Stub*. See text for references to published sequences.

of the conserved regions share sequence similarity with the active sites of two fungal ribonucleases, and the other three contain mostly hydrophobic residues.

The pairwise amino acid sequence identities were estimated from the aligned sequences (Table I) and they reveal a remarkably wide range of sequence variation (38.7% to 93.5% identity). The most divergent pair of S-alleles (S_{2nic} and $PS2A$) differs at 132 amino acid positions, while the most similar sequences (S_{2pet} and $PS2A$) differ at 13 amino acid positions. The average amino acid identity among pairs of *Nicotiana* alleles is 53.18%, and the average divergence among all interspecific allele comparisons is 50.56%. There are many cases in the distance matrix in which a pair of alleles from one species appears more divergent than alleles from different species. This observation led to the hypothesis that the S-alleles exhibit shared polymorphisms (Ioerger *et al.*, 1990; Clark and Kao, 1991), discussed in greater detail in a subsequent section of this chapter. A neighbor-joining tree (Saitou and Nei, 1987) was constructed from the distance matrix of Table I, and it exhibits the clustering of alleles from different species (Fig. 2). This tree was found to be robust with respect to several methods of building the distance matrix, including correction for multiple hits or the use of of the PAM250 matrix of Dayhoff to correct for the inferred number of nucleotide substitutions necessary for various amino acid changes. In both cases the tree topology was unchanged. Confidence in each of the nodes of the tree was assessed by a bootstrap analysis. One thousand trees were constructed from bootstrap samples of polymorphic sites, and the percentage of resulting trees that had each observed node is printed. The unusually high bootstrap confidence percentages are attributable to the extraordinary magnitude of divergence among these sequences.

The sequence comparisons of S-alleles pose two evolutionary puzzles, including how the sequences can be so very divergent and how the structure of the gene genealogy, which seems to reflect allele sharing, could arise. The approaches of both classical population genetics and of modern molecular evolution contribute to our understanding of these problems.

DETERMINISTIC POPULATION GENETICS THEORY FOR A
SELF-INCOMPATIBILITY LOCUS

A brief review of the classical population genetics of self-incompatibility loci will provide a theoretical underpinning and make it clear why these loci are so diverse. Suppose there are initially three S-alleles segregating in a population, and that these alleles have no epistatic effects that influence fitness. If the frequencies of alleles S_1, S_2, and S_3 are p_1, p_2, and p_3, and the genotype frequencies are P_{ij} for $S_i S_j$, then the recurrence equations can be written

$$P'_{12} = 1/2(1 - P_{12})$$
$$P'_{13} = 1/2(1 - P_{13})$$
$$P'_{23} = 1/2(1 - P_{23}).$$

Regardless of the initial frequencies of the alleles, this system will evolve to the point $p_1 = p_2 = p_3 = 1/3$. When trajectories of the allele frequency changes are plotted on a de Finetti diagram, they all describe straight lines through the centroid. Allele frequencies do not always change monotonically toward the equilibrium, however, and can increase from very rare to exceed the equilibrium value in a single generation. This is readily seen from the recurrence equations for alleles frequencies:

$$p'_1 = 1/2 p_1 + 1/2 P_{23}$$
$$p'_2 = 1/2 p_2 + 1/2 P_{13}$$
$$p'_3 = 1/2 p_3 + 1/2 P_{12}.$$

If S_1 is initially very rare, so that $P_{23} \approx 1$, then the following generation the frequency of S_1 will be nearly $1/2$ (which exceeds the equilibrium value of $1/3$). Note that no allele frequency can ever exceed a frequency of $1/2$, because all individuals in the population must be heterozygous. This places an important boundary condition on diffusion approximations for the evolution of self-incompatibility alleles.

   If n distinct alleles are segregating in the population, then it is somewhat more cumbersome to express the recurrence equations, but Nagylaki's (1975) notation is the clearest:

$$P_{ij} = 1/2 \sum_{k \neq i, j} \left[\frac{P_{ik} P_j}{1 - p_i - p_k} + \frac{P_{jk} P_i}{1 - p_k - p_j} \right],$$

where the sum is over alleles 1 through n excluding alleles i and j. Apparently the first demonstration that $p_i = 1/n$ is a stationary point of this recursion was due to Moran (1962), and the proof of its local stability was given by Nagylaki (1975). Numerical studies indicate that the symmetric equilibrium is unique and globally stable (Ewens and Ewens, 1966). Allele frequencies do not always change monotonically toward the equilibrium. For example, if there are four common alleles, with frequencies 0.2, 0.3, 0.25, 0.25, and a fifth very rare allele, then as the fifth allele begins to increase in frequency, the other four alleles will first approach the four-allele equilibrium. This means that the allele whose

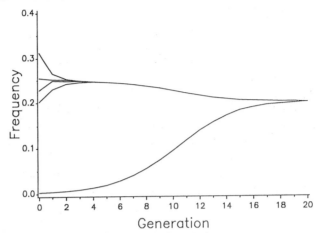

Fig. 3. Trajectories of allele frequencies for the deterministic gametophytic self-incompatibility model (Nagylaki, 1975). Initially the four common alleles converge to a frequency near 0.25, then, as the rare fifth allele increases in frequency, all five alleles converge to an equilibrium frequency of 0.2. Note the rapid approach to equilibrium.

initial frequency was 0.2 will first increase in frequency to nearly 0.25, then, as the fifth allele increases in frequency, all four of the original alleles will decrease in concert to a new equilibrium of 0.2 (Fig. 3).

Referring back to the situation with three equally frequent alleles, note that an S_1 pollen grain can only be successful in effecting pollination on the $S_2 S_3$ plants, which represent 1/3 of the population. Suppose that a novel mutation occurs that generates a fourth functional allele. Pollen with the new allele can successfully pollinate any plant in the population, giving it a 3-fold fitness advantage over the three extant alleles. In general, if there are n alleles, the $n+1$st allele will have a fitness advantage of $n/(n-2)$ assuming the population is at the equilibrium with all alleles equally frequent. Even if there are 10 alleles in the population, the rare novel mutation has a 25% advantage over the extant alleles. The S-locus will depart markedly from a neutral locus in that novel mutations always have an advantage, and are much more likely than a neutral allele to increase in frequency. An important theoretical question that can be put to empirical test is to describe the balance between the influx of new mutations and their loss through genetic drift. The deterministic theory already suggests that the S-locus will have far more alleles segregating than will a neutral locus.

THE NUMBER OF ALLELES MAINTAINED IN A FINITE POPULATION

The seminal paper of Wright (1939) began a long controversy with R.A. Fisher about the number of self-incompatibility alleles that could be maintained by mutation in a finite population. Both Wright and Fisher applied diffusion theory, and a number of potentially troublesome assumptions and approximations had to be made to do this. Two assumptions of the diffusion approach were violated by a self-incompatibility locus (Moran, 1962). First, the expected mean change in allele frequency $[M(x)]$ and its variance $[V(x)]$ must be of order $1/N$ before application of the diffusion is appropriate (Ewens, 1964). If the number of alleles is below the equilibrium number, a rare allele has an enormous expected mean change in frequency. Secondly, it is assumed that the mean change in allele frequency is independent of the frequency and number of other alleles in the population. For a neutral gene, the change in frequency of an allele does not depend on the frequencies of other alleles, but for self-incompatibility alleles, allele frequency changes do exhibit this dependence (see Fig. 3 for an example). Although this violation of the Markovian property led investigators to question Wright's application of diffusion theory, in practice, the approximation can still be quite good. The quality of the approximation could not be known *a priori*, however, and we are only happy with it now because of computer simulations demonstrating its accuracy (Kimura, 1965; Ewens and Ewens, 1966; Mayo, 1966).

Because the mean change in allele frequency depends on the frequencies of the other alleles in the population, we can obtain an expression for $M(x)$ if we assume that one allele has frequency x, and the remaining $k-1$ alleles each have frequency $(1-x)/(k-1)$. Wright (1939) obtained an expression for the change in allele frequency as $\Delta q = q(q-\hat{q})/(1-3\hat{q}+2q\hat{q})$, which after substituting $x=q$ and $1/k=\hat{q}$, gives (ignoring mutation):

$$M(x)=\frac{x(1-kx)}{(k-3-2x)}.$$

Fisher (1958) obtained the expression $p(p-\alpha)/(1-2\alpha)$ for the mean change in allele frequency per generation, and substituting $x=p$ and $1/k=\alpha$, this yields

$$M(x)=\frac{x(1-kx)}{(k-2)}.$$

These two expressions for the expected change in allele frequency fall to zero at $x=1/k$, as they should because this is the deterministic equilibrium fre-

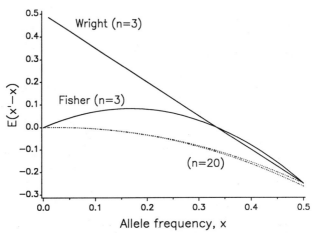

Fig. 4. Wright and Fisher used different approximation to the mean change in allele frequency per generation. This quantity, written as Δq, $E(x' - x)$, or $M(x)$, must be approximated because it depends on the frequencies of all alleles. For $n=3$ alleles, Wright's approximation was much more accurate, but with 20 alleles the curves are virtually identical.

quency, and for $x < 1/k$, both are positive, so the frequency of rare alleles increases. Although no population has been found with only three alleles, the expressions for $M(x)$ differ dramatically when $k=3$ (Fig. 4). When one allele is exceedingly rare, the next generation its allele frequency will be nearly 1/2, so Wright's formula is more accurate in this case. In the case of a more biologically reasonable number of alleles (*e.g.*, $k=20$), the two approximations are very close. Although Wright initially used the variance term $V(x) = q(1-q)/2N$, which is the variance in the change in allele frequency for a neutral allele, Fisher (1958) objected and gave the exact variance for the self-incompatibility allele as

$$V(x) = x(1-2x)/2N.$$

Fisher (1958) went on to claim that his results departed markedly from those of Wright. Wright (1960, 1964, 1969) conceded that Fisher's formulation of the variance was more accurate, but he convincingly showed that the two approximations of the allele frequency distribution differed only negligibly. Figure 4 also demonstrates the enormous strength of selection in changing allele frequencies that deviate widely from the equilibrium, and it shows that the advantage to a very rare allele is greater when there are fewer alleles.

Substituting the values of $M(x)$ and $V(x)$ into Wright's (1938) formula for the stationary distribution, we get

$$\phi(x) = 4Nu \ e^{2Neax}(1-2x)^{2Neb-1}/x,$$

for $0 < x < 1/2$, where $a = 1/[(1-F)(1-2F)]$, $b = u + 1/[2(1-F)]$, and $F = \sum_i x_i^2$ (Yokoyama and Nei, 1979). For a given population size and mutation rate, we can numerically solve the expected number of alleles ($n_e = 1/F$) at equilibrium (see below), and use this in numerical solutions of $\phi(x)$. Figure 5A

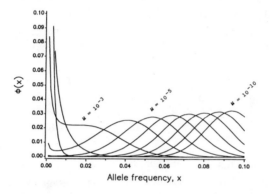

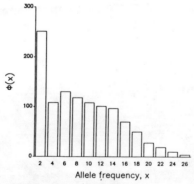

Fig. 5. A: $\phi(x)$, the expected number of alleles whose frequency is x, for a population size of 500 and a range of mutation rates from 10^{-1} to 10^{-10}. At low mutation rates, the allele frequencies cluster around the equilibrium value of $1/k$, and $\phi(x)$ has very little density at low frequencies, because of the rapid increase in frequency when alleles are rare. As the mutation rate increases, the mode of $\phi(x)$ decreases, and a spike of very rare, newly introduced alleles appears. B: results of simulations for a mutation rate of 10^{-3} and a population size of 500. Allele frequencies are expressed as counts.

shows plots of $\phi(x)$ for $N=500$ and a range of mutation rates. At very high mutation rates there is rapid allele turnover and $\phi(x)$ is monotonously decreasing, as it is for a neutral gene. As the mutation rate decreases, the number of alleles decreases, and $\phi(x)$ has increasing density at higher allele frequencies. When few alleles are maintained in the population, the average allele frequency is quite high because of the evenness of the allele frequency distribution(recall that the deterministic equilibrium allele frequency is $1/k$ for all alleles). As the mutation rate decreases further, the mean of $\phi(x)$ shifts to the right as the number of alleles decreases, and the variance of $\phi(x)$ decreases. Simulation of a population with $N=500$ and $u=10^{-3}$ produces a mean $\phi(x)$ shown in Fig. 5B, which agrees well with the diffusion result of Fig. 5A. More extensive numerical simulations of Ewens and Ewens (1966) and of Mayo (1966) also confirmed the quality of the diffusion approximation.

Wright (1939) obtained expressions for the number of S-alleles maintained in a finite population by equating the rate of influx of new alleles to the rate of allele loss. A simpler approach is to note that

$$\int_{1/2N}^{1} x\phi(x)dx = 1$$

and to solve for the effective number of alleles, $n_e = 1/F$ (Yokoyama and Nei, 1979). Figure 6 shows the number of alleles maintained at an S-locus over a range of population sizes and mutation rates. The numerical simulations described above demonstrate the accuracy of the approximations made in deriving these results (Fig. 7).

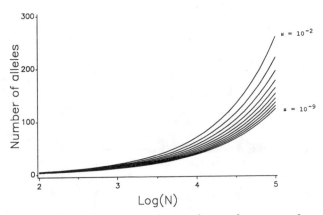

Fig. 6. The number of S-alleles maintained at steady state for a range of population sizes and mutation rates. The figures are derived from Eg. 36 of Yokoyama and Nei (1979).

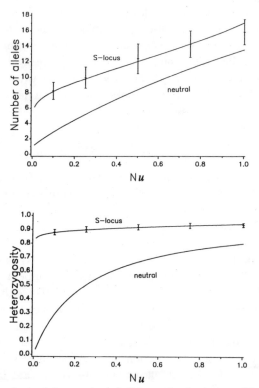

Fig. 7. Heterozygosity and the number of alleles maintained at an *S*-locus and a neutral locus in a finite population. Results for the neutral locus are from the infinite alleles model of Kimura and Crow (1964).

Wright's (1939) motivation for the above analysis was to determine whether the mechanism of self-incompatibility could explain Emerson's observation of 45 alleles in a population of perhaps 500 Oenothera. He pointed out that a mutation rate of 0.001 is required to obtain 40 alleles in a population of 500 individuals, and this mutation rate is untenable because Emerson detected no mutations in 45,000 pollen. Wright showed that extreme population subdivision, as might occur if only very near neighbors can pollinate one another, might yield the observed number of alleles with acceptable mutation rates. (See also Takahata's chapter in this volume for a demonstration that symmetric overdominance in a finite subdivided population can retain many more alleles than can a panmictic population.) Fisher (1958) felt that a more likely explanation was that the population had recently decreased in size from perhaps 4,000 to 500 individuals. Such a bottleneck would result in a loss of fewer alleles and fewer segregating sites for a self-incompatibility locus than a neutral locus (Watterson, 1984; Tajima, 1989) because of the relative evenness of the S-allele frequency distribution. The controversy has never really been settled, but Yokoyama and Hetherington (1982) added another twist to the story by examining the distribution of the number of S-alleles in a sample from a population. Using the expression

$$n_s = \int_{1/2N}^{1} [1-(1-x)^m] \phi(x)dx$$

they could solve the expected number of alleles in a sample of size m from a population of size N having mutation rate u. Of course, the number of alleles in a sample must be less than or equal to the number of alleles in the population. Nevertheless, they were able to obtain reasonable fits to Emerson's data with a population size of 2,000 and mutation rate of 10^{-5}.

Another important feature of S-gene evolution that Wright (1939) considered was the rate of allele turnover. He used the argument that the rate of influx of new alleles by mutation and selection must balance the the rate of loss in order to calculate the expected number of alleles at steady state. If we define L as the rate of loss of alleles at a neutral locus, then Wright found the rate of loss of S-alleles to be approximately $L(n-3)/(n-1)$. A large population with few alleles will result in strong selection favoring new alleles and make it very improbable to lose any extant alleles. The turnover rate increases as the number of alleles increases, and approaches the neutral rate in populations with $Nu \gg 1$ (see section on coalescence). Maruyama and Nei (1981) showed that the substitution rate was greater for an overdominant locus than for a neutral locus because new mutations are initially found in heterozygotes and thus have a selective advantage. The substitution rate approaches that of a neutral locus when the heterozygosity is very high, however, because this selective advantage decreases with increasing population heterozygosity. Simulations of Takahata and Nei (1990) support the finding that overdominant selection accelerates the rate of substitution. These same findings have been confirmed for an S-locus.

SHARED POLYMORPHISM OF S-ALLELES

The phylogenetic relationships among the S-alleles depicted in Fig. 2 suggest that the polymorphisms are more ancient than the times of species divergence. Although there are no clear cases of an entire allele being shared by two different species, the species have been genetically isolated long enough that the accumulation of mutations since the speciation might have erased such a match. Instead, we might expect that the particular sites that are involved in the determination of self might still be segregating. What is needed is a means to examine the sites that are segregating for the same pair (or more) of different nucleotides (or amino acids), and to test the statistical significance of these "shared polymorphic" sites. Such sites may exhibit shared polymorphism either because of the common ancestry with two different alleles, or the shared polymorphism may have arisen by chance in the two species after they diverged.

We constructed that null distribution of the number of shared polymorphic sites that would arise by chance by generating samples on the computer having the same number of segregating sites as the observed data, but with random positions of the polymorphisms (Ioerger *et al.*, 1990). Because the observed number of shared polymorphic sites was found to be greater than any of the 1,000 computer samples, the data suggest that shared polymorphism occurs more frequently than chance alone predicts. The computer resampling may underestimate the number of shared polymorphisms that occur by chance if selective constraints concentrate the polymorphism on a restricted set of sites. The resampling was repeated to reflect the extreme situation of polymorphism restricted only to those sites actually observed to be polymorphic in the sample. This resulted in a greater average number of shared polymorphisms, but every such random sample had fewer shared polymorphisms than were actually observed. Similarly, when the analysis was repeated with amino acid polymorphisms, an excess of shared amino acid polymorphism was found (Ioerger *et al.*, 1990). The excess of shared polymorphisms is consistent with the idea that the alleles are unusually old, because the common ancestor of *Nicotiana* and *Petunia* is thought to have lived approximately 40 million years ago (Wolfe *et al.*, 1989).

The method of Nei and Gojobori (1986) or Li *et al.* (1985) can be used to estimate the rates of synonymous and nonsynonymous substitution. There are several important assumptions that are implicit in the approach of inferring natural selection from elevated nonsynonymous substitution rates. First, it is assumed that the rates of mutation are homogeneous, and that the synonymous substitution rate is close to the neutral mutation rate. If natural selection favors amino acid substitutions at many sites, and each substitution is sufficient to affect fitness, then, assuming that the respective mutation rates are nearly the same, we expect that the synonymous substitution rate will not be greatly affected, while the nonsynonymous substitution rate will be increased. Clark

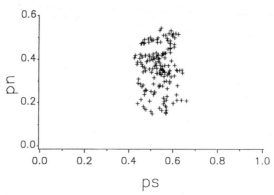

Fig. 8.   Proportion of synonymous (p_s) and nonsynonymous (p_n) substitutions among the 12 S-allele nucleotide sequences described in Clark and Kao (1991). A window of 20 codons was swept along the gene, and for each position, p_s and p_n were calculated.

and Kao (1991) found that the synonymous substitution rate was homogene-ous across the S-locus, while there was significant heterogeneity in the non-synonymous substitution rate. Many genes exhibit a correlation between p_s and p_n (Graur, 1985; Shields *et al.*, 1988), but the homogeneity of p_s for the S-locus departs from this pattern (Fig. 8).

The very low mutation rate of S-alleles, and the fact that all S-alleles examined differ by several (or many) amino acid differences, suggests that more than one substitution is necessary to determine a new S-allele. This implies that the first nonsynonymous mutation may be neutral, so that several synonymous mutations may occur before a subsequent nonsynonymous mutation occurs to generate a new functional allele. Selection would then have the effect of increasing both the synonymous and nonsynonymous rates, and the amplifica-tion of the nonsynonymous rate would be much lower than the case in which selection favors any single amino acid change. Similarly, if only a small subset of amino acid sites are subjected to selection, there may be very low statistical power in detecting selection by examining the rates of synonymous and nonsynonymous substitution averaged over the entire gene. Clark and Kao (1991) avoided these problems by partitioning the sites based on whether they exhibited shared polymorphism or not, and then compared rates of synony-mous and nonsynonymous substitution. The result was that shared polymor-phic sites, which are more likely to be involved in determining functional differences, do show an excess of nonsynonymous substitution. The idea that polymorphisms may be retained in a population since the common ancestor of

Solanaceae (about 40 MY) is only made plausible after a consideration of the expected coalescence time of S-alleles.

COALESCENT PROPERTIES OF SELF-INCOMPATIBILITY ALLELES

The availability of molecular sequence data, which allows not only ascertainment of allele identity but also the degree of divergence between alleles, provides compelling motivation to understand the coalescence properties of S-alleles. The properties of the genealogy of neutral alleles are well understood (Kingman, 1982; Griffiths, 1980; Tavaré, 1984; Watterson, 1984), and the mathematics are tractable because a neutral gene can be handled as a one-dimensional Markov chain. The complete analysis of coalescence models in the face of natural selection requires solution to a multidimensional diffusion, a forbiddingly complex problem for which little progress has been made. In the case of strong selection and weak mutation, Gillespie (1989) showed that his fluctuating selection model produced a gene genealogy whose distribution of coalescence times had the same structure as a neutral gene. Takahata (1990a, 1991) stresses the distinction between a gene genealogy, which is based on a random sample of genes, and an allele genealogy, which is obtained when functionally distinct alleles are characterized. In the case of the major histocompatibility complex (MHC) loci and the S-locus, the first sequences that are gathered come from distinct alleles, and in both cases the marked divergence in the allele genealogies are consistent with strong selection operating. Under a model with strong symmetric balancing selection, in which all homozygotes have fitness $1-s$ and all heterozygotes have fitness 1, Takahata (1990b) demonstrated a remarkable property of the allele genealogy. He first obtained an expression for the rate of allelic turnover by applying the diffusion approximation. Following Crow and Kimura (1970), he obtained an expression for the expected time, $t(x)$, for an allele of frequency x to be lost. He applied the condition of strong selection in assuming that once an allele attained a frequency δ, it was bound to increase to a common equilibrium frequency near the deterministic equilibrium. The expected time that an allele whose initial frequency is $x < \delta$ becomes common is $t^*(x)$. At equilibrium these two quantities must be equal (no net flux in allele frequency), and the rate is the rate of allelic turnover. The form of selection at the S-locus can be considered extreme overdominance, with all homozygotes lethal and all heterozygotes of equal fitness (*i.e.*, $s = 1$). The $M(x)$ term that Takahata (1990b) used was (letting $s = 1$ and ignoring mutation) $x(x - F)/(1 - F)$, which produces a smaller allele frequency change than does Wright's $M(x) = x(x - F)/(1 - 3F + 2xF)$. We will examine the exact consequences of this difference elsewhere, but for now assume that these will produce much the same results near equilibrium provided many alleles are segregating. The time since the most recent allelic turnover is exponentially distributed, and Takahata (1990b) derived an expression for the probability that a pair of alleles in a sample diverged exactly K allelic turnovers ago. From this he obtained an exponential distribution for the expected coalescence time of a sample of alleles, with a rate parameter depending on the rate of allelic turnover. The remarkable result is that the genealogy of alleles under strong balancing selection has the same structure as alleles at a neutral locus, but the time scale of the genealogy under selection was considerably expanded. By comparing the turnover rates for the neutral and selected case, Takahata (1990b) found that the allelic genealogy could be scaled by a factor f_s, which in the case of an S-locus, where $s = 1$, is

$$f_s = \sqrt{2N}/2Nu[\log(1/8\pi Nu^2)]^{-3/2}.$$

Whereas the expected time to coalescence for a pair of neutral alleles is $2N$ generations, the expected time to coalescence for a pair of S-alleles is $2Nf_s$ generations. By similar reasoning, the substitution rate for an S-locus relative to that of a neutral gene is found to be

$$\alpha = (\sqrt{2}/2) \log[1/(8\pi Nu^2)].$$

Although the allelic genealogy in a steady-state population has the same structure as a neutral gene, the structure of the genealogy will differ markedly from a neutral locus if the population is not at equilibrium. For example, if there are fewer alleles in the population than the equilibrium, the turnover rate will be much lower than at equilibrium.

As was described by Maruyama and Nei (1981), the rate of substitution in the model of symmetric overdominance decreases as Nu increases. This is because, for a fixed population size, a higher mutation rate results in a greater number of alleles, which results in a weaker selective advantage to any newly introduced allele. The same principle applies to the S-locus. Populations that have, many alleles will have mostly compatible matings, so the difference in fitness between extant and new rate alleles will be very small. Thus, for a given population size, as the mutation rate increases, the substitution rate approaches the neutral substitution rate. Relative to a neutral gene, however, when the mutation rate is low, a new S-allele mutation is strongly favored, so the substitution rate is much higher than that of a neutral gene.

A series of Monte Carlo simulations were performed to determine the coalescence times for S-alleles. Each generation N heterozygous genotypes were chosen, and random numbers were picked to determine whether each gamete mutated. If a mutation occurred, it generated a new allele. Each allele was associated with two vectors, one that kept track of the ancestral history of the allele and another the times that mutations occurred, in the manner of Takahata and Nei (1990). The population size was fixed at 100 individuals, and mutation rates of 10^{-5}, 10^{-4}, 10^{-3}, and 10^{-2} were simulated in 100 replicate populations for $400N$ generations. The results of the simulations, reported in Table II, are in good agreement with the results of Takahata and Nei (1990) for the case $Ns = 100$. Representative neutral and S-allele genealogies are shown in Fig. 9. If the mutation rate were increased, the neutral tree would gain more alleles, but the expected time to coalescence of all the genes would remain $4N$ generations. The S-allele genealogy, on the other hand, would decrease in coalescence time as the mutation rate increases. As the mutation rate increases, the S-allele genealogy would appear more and more like the neutral genealogy. Table II also shows that the expected time to coalescence of

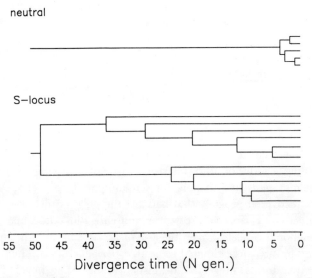

Fig. 9. Gene genealogy for a neutral gene and for a representative S-locus Monte Carlo simulation. In both cases, $N = 100$ and $u = 10^{-4}$. If the mutation rate were increased, the neutral tree would increase in allele number, but the expected coalescence time would not change, whereas the S-locus would gain alleles and the coalescence time would decrease. As described in the text, the topology of these trees is the same, but the time scales differ dramatically.

TABLE II

Results of Monte Carlo Simulations of S-locus Evolution

Nu	u	H	n_a	T_c	T_d	$2f_s$
0.004	10^{-5}	0.822 ± 0.021	6.60 ± 0.67	344.17 ± 86.21	231.85 ± 91.80	238.72
0.04	10^{-4}	0.847 ± 0.018	8.10 ± 0.86	87.68 ± 31.25	60.96 ± 22.11	41.02
0.4	10^{-3}	0.880 ± 0.015	11.44 ± 1.68	37.41 ± 14.29	21.14 ± 12.64	9.68
4.0	10^{-2}	0.928 ± 0.011	24.68 ± 3.16	10.16 ± 5.93	8.82 ± 4.60	8.72

100 replicate populations of size $N = 100$ followed for 40,000 generations. Reported are the mean heterozygosity $(1-F) \pm 1$ standard deviation, the mean number of alleles, the mean time to coalescence of alleles in the population (T_c), and the mean pairwise divergence time (T_d). T_c and T_d are reported in units of N generations. $2f_s$ is the theoretical prediction for the mean coalescence time for pairs of alleles (in units of N generations).

a pair of alleles ($2f_s$ in units of N generations) is in very good agreement with the numerical simulations.

Genes in a sample that fall within the same functional allele also have a gene genealogy that is identical to that of a neutral gene. N for genes within an allele will be reduced, so their sequence variation and time to coalescence will be similarly reduced. If a sample of genes were taken from a natural population, we would expect the structure to look like the S-allele genealogy of Fig. 9 with each major allelic lineage having a neutral genealogy at its tip. The sequence divergence within an allelic class is expected to be far less than the divergence between alleles.

ORIGIN OF NOVEL S-ALLELES

The rate at which new S-alleles are introduced into the population by mutation is a critical parameter in the models described above. Lewis (1948, 1949, 1951) did a series of mutagenesis experiments, treating self-incompatible plants with mutagens and looking for self-compatible progeny. Although he successfully recovered progeny that had lost the pollen function of the S-allele (as determined by reciprocal crosses), in no case did one S-allele mutate to another functional S-allele. Given the size of Lewis' samples, the upper bound in the mutation rate was estimated to be 10^{-8} per generation. The nature of the pollen inactive forms suggested to Fisher (1961) that novel S-alleles are normally generated by crossingover, and the pollen-inactive forms are aberrant recombinants possessing antigenic properties of both parental alleles. The problem with Fisher's model was that it predicts that S-alleles would fall into classes of mutually self-compatible groups, and Lewis (1962) pointed out that S-alleles do not behave this way. Empirical studies of recombination among

S-alleles have not been sufficiently large to fully address this issue, but the experiments of Lewis (1949) failed to find recombinant products of active and inactive S-alleles. While his experiments show that the S-system is not composed of two loosely linked loci, one controlling specificity and one controlling a rejection reaction, neither do the results prove that both functions are encoded by a single gene.

The molecular sequence data shows that alleles differ from each other at any amino acid sites. The most similar pair of alleles that have been sequenced are *S2pet* and *PS2A* of *P. inflata* and *P. hybrida*, respectively, and they differ at 13 amino acid sites. Unfortunately, given the way alleles have been sampled (based on function) and the small number of alleles examined within any one species, this cannot be taken as evidence that a single amino acid change is not sufficient to specify a new functional allele. But the extraordinary sequence divergence and the unusually low mutation rate are consistent with the requirement for more than one mutation to change allelic identity.

One means of generating new alleles when multiple substitutions are necessary is intragenic recombination. Some of the diversity of MHC alleles is generated in part by intraexonic recombination (She *et al.*, 1991; Gyllensten *et al.*, 1991), which may help to generate and maintain the high levels of diversity at these loci. Ebert *et al.* (1989) and Kaufman *et al.* (1991) suggested that gene conversion may be responsible for the generation of new S-alleles, but there now appears to be little basis for this idea. One intriguing experimental result is that although mutagenesis is apparently incapable of generating new functional alleles, close inbreeding of self-incompatible plants can result in appearance of new S-alleles (Lewis, 1948). It is critical to obtain the sequence of those alleles that were generated by inbreeding, because if they are intragenic recombinants, their origin would be recent enough that subsequent substitutions would not mask the appearance of the recombination event in the S-allele DNA sequence.

Linkage disequilibrium among neutral genes decays by recombination at a rate that depends on the frequency of recombination. There is an extensive literature on the effects of natural selection and drift on linkage disequilibrium (see Karlin, 1975; Hudson, 1985), and a general conclusion is that tighter linkage will allow greater linkage disequilibrium. The 12 alleles that were examined in the Clark and Kao (1991) study were examined for pairwise linkage disequilibria among nucleotide sites. There were a total of 395 polymorphic nucleotide sites, and the Fisher exact test was performed on all 77,815 possible pairwise comparisons. We found 5,857 of them were significant at the 5% level (this is 7.5% of all tests), and 1,584 were significant at the 1% level (or

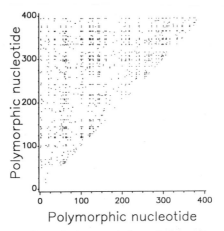

Fig. 10. Fisher exact test for pairwise associations (linkage disequilibrium) among nucleotides of the 12 S-allele sequences examined in Clark and Kao (1991). Only the tests that were significant at the 1% level were plotted.

nearly twice as many as expected by chance). Given that alleles from different species were included in this analysis, it is striking that there were not more significant disequilibria. The pairs of sites that exhibited significant association showed no particular pattern across the gene (Fig. 10). By themselves, these results neither support nor disprove the possibility for intragenic exchange.

There are two lines of evidence suggesting that there is little intragenic recombination among S-alleles. The first entails statistical tests of clustering of the polymorphic sites within the gene. The tests of Sawyer (1989) and Stephens (1985) are very sensitive to intragenic exchange, and they gain statistical power with larger numbers of segregating sites. Both tests can confuse recombination with cold spots of polymorphism caused by purifying selection, and a means of avoiding this problem is to examine only silent polymorphisms. Both tests fail to reject the null hypothesis that the segregating sites accumulated in the absence of recombination (Clark and Kao, 1991). These results were viewed as being consistent with the pattern revealed by the genomic dot plot, which shows complete loss of homology flanking the S-locus. Such a pattern could only be obtained in the absence of both selective constraints and recombination, and it suggests that there is a mechanism for preventing chromosome pairing in the region of the S-locus. Although the above analyses consistently indicate that there is little if any intragenic recombination among extant S-alleles, recall that the alleles are extremely old, and evidence for exchange in the distant past could have been obliterated by subsequent mutations.

MODIFIER MODEL FOR THE EVOLUTION OF INTRAGENIC RECOMBINATION RATE

The evolution of intragenic recombination was explored with a model that allows the rate of intragenic recombination to be regulated by a modifier locus. In this model, the S-locus behaves as described above, with mutation rate u to new functional alleles. Linked to the S-locus with a recombination rate r is the modifier locus. The genotypes MM, Mm, and mm at the modifier locus have rates of intragenic recombination within the S-locus of r_1, r_2, r_3. If an intragenic recombination event occurs, it produces a new functional S-allele or a completely non-functional allele in the proportions $a : 1 - a$. Intragenic recombination, like mutation, may result in a loss of normal function of either the pollen or the style. Mutant pollen may be accepted by no other genotype (sterile), or may be accepted by all genotypes (self-compatible). Similarly, stylar function may be altered to allow no pollen to germinate (sterile) or any pollen to germinate (self-compatible). Initially we let $1 - a$ be the fraction of recombinant S-alleles that are sterile in both functions. If $a = 1$, every intragenic recombination event produces a novel S-allele, so the influx of new alleles is increased as it would be if the mutation rate were increased. This is illustrated by Fig. 11,

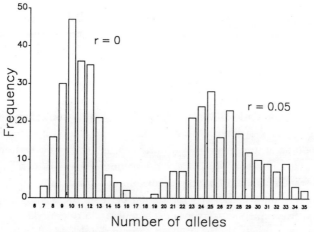

Fig. 11. Number of alleles maintained at an S-locus with intragenic recombination. 200 replicate populations of size $N = 200$ with $u = 0.001$ were simulated. The histogram on the left represents the case with no intragenic recombination, and the histogram of the right represents populations with an intragenic recombination rate of 0.05, where every recombination event generates functionally novel S-alleles.

TABLE III

Invasion of Intragenic Recombination Modifier Alleles

r_1	r_2	No. of cases of r_2 invasion in 100 reps
0	0	0
	0.0001	2
	0.005	2
	0.01	8
0.0001	0	1
	0.0001	3
	0.005	6
	0.01	5
0.005	0	2
	0.0001	0
	0.005	2
	0.01	4
0.01	0	1
	0.0001	0
	0.005	0
	0.01	2

Numerical simulations of a two locus model were performed with $N = 50$, $u = 0.001$, and the rate of recombination between the S-locus and the modifier, $r = 0$. r_1 and r_2 are the rates of intragenic recombination within the S locus for genotypes MM and Mm, respectively.

in which the distributions of numbers of alleles in 200 replicate populations with $N = 50$ and $u = 0.005$ are shown for the case of zero intragenic recombination and with $r_1 = 0.05$. Intragenic recombination increases the equilibrium number of S-alleles as suggested by Fisher (1961), and when $a \approx 1$, the effect of intragenic recombination on the heterozygosity and number of alleles at an S-locus is much greater than for a neutral locus (Hudson, 1983).

Numerical simulations were performed to identify conditions for invasion of modifier alleles that alter the rate of intragenic recombination. After sampling a genotype, a uniform random number was drawn to determine whether mutation occurs, a second uniform random number was drawn to determine whether intragenic recombination occurs, and a third random number was drawn to determine whether recombination between the S and M loci occurs. This process was repeated $2N$ times each generation to generate the sample of $2N$ haplotypes. Initially the population is fixed for the M allele with specified intragenic recombination rate r_1. After $1/u$ generations, the m allele is introduced at a frequency of $1/2N$ in a random S-haplotype, and Mm genotypes produce intragenic recombinant gametes at rate r_2. The fate of the m allele is then followed until it is either lost or another $1/u$ generations have passed. The results of the simulations are given in Table III for the case $a=1$. If $a=1$, so that every recombinant is functional, we found that the modifier allele is more likely to invade than a neutral allele if it results in increased intragenic recombination ($r_2 > r_1$). If $r_2 < r_1$, invasion occurs only due to drift, and much less frequency. If $a \ll 1$, so that most intragenic recombinants are sterile, the modifier invades only if it decreases the rate of intragenic recombination (results not shown).

CONCLUSIONS

Five points concerning the evolution of self-incompatibility alleles warrant special emphasis. First, as has been recognized since the 1930s, even small populations of gametophytically self-incompatible plants have a large number of S-alleles, despite the apparently low mutation rate. The number of alleles that is maintained in natural populations appears roughly concordant with the prediction of classical population genetics theory, but molecular variation in samples from natural populations is needed to more fully test the models. Second, sequence analysis of the alleles has shown that they exhibit extraordinary diversity. This diversity is consistent with the maintenance of polymorphism in the population for very long periods of time, allowing nucleotide substitutions to accumulate differences between alleles. The third point, which is not predicted by classical population genetics, is that the alleles exhibit an excess of shared polymorphism. Observation of shared polymorphism is consistent with polymorphisms having been present in the ancestor to extant species. Consideration of the distribution of coalescence times for self-incompatibility alleles has produced the fourth observation, that S-alleles are expected to exhibit the extraordinary diversity that they do. The theory emphasizes the distinction between functionally different alleles (whose common ancestor is generally very ancient) and independent samples of the same functional allele (which are predicted to be more similar than neutral alleles). Many aspects of the patterns of S-allele polymorphism within and between species are broadly consistent with population genetic theory, but more sequence data are needed before it will be possible to provide a true test of the coalescence models.

SUMMARY

Self-incompatibility in solanaceous plants is determined by a major locus that encodes the S-protein. In gametophytic self-incompatibility, S-protein expression results in successful pollination only when the pollen allele differs from both maternal alleles. The S-locus has been cloned and sequenced in several species, and it exhibits extraordinary intraspecific sequence diversity. Analyses of the distribution of the number of shared polymorphic sites verifies that the sequence diversity is due to the extreme age of the alleles. Despite the strong natural selection that is acting on this locus and the initial advantage given to a rare functionally distinct allele, the gene as a whole exhibits an excess of synonymous substitutions. Aspects of the mutational and recombinational history of the locus can be inferred from the sequence variation. New alleles appear to arise primarily by accumulation of point mutations, and there is as yet no evidence for intragenic recombination generating new alleles. Classical population genetic theory correctly predicts that an S-locus should exhibit a much larger number of alleles than a neutral locus, and the allele sequence data allows inferences to be drawn from the gene genealogy of S-alleles. The coalescence properties of an S-locus also dictate that the expected time to coalescence and the expected number of sites that differ between a pair of alleles is much greater than that of a neutral locus. The degree of inflation of the depth of an S-allele tree over that of a neutral gene depends on Nu^2, where N is the population size and u is the rate of mutation to new distinct S-alleles. In fitting coalescent models to the data, it is important to consider both the genealogy of all sampled genes and the genealogy of functionally distinct alleles. Only an allele genealogy is available at present, and we will make predictions about the properties of the gene genealogy using population genetic principles.

Acknowledgments

I thank Teh-Hui Kao for introducing this problem to me, and for sharing the data and insights that he and his students have gathered. This work was supported by a grant from the U.S. National Science Foundation and a sabbatical award from the Alfred P. Sloan Foundation.

REFERENCES

Ai, Y., E. Kron, and T.-H. Kao, 1991. *Mol. Gen. Genet.* **230**: 353–358.
Ai, Y., A. Singh, C.E. Coleman, T.R. Ioerger, A. Kheyr-Pour, and T.-H. Kao, 1990. *Sex. Plant Reprod.* **3**: 130–138.
Anderson, M.A., E.C. Cornish, S.L. Mau, E.G. Williams, R. Hoggart, A. Atkinson, I. Bonig, B. Grego, R. Simpson, P.J. Roche, J.D. Haley, J.D. Penschow, H.D. Niall, G.W. Tregear, J.P. Coglan, R.J. Crawford, and A.E. Clarke, 1986. *Nature* **321**: 38–44.
Anderson, M.A., G.I. McFadden, R. Bernatzky, A. Atkinson, T. Orpin, H. Dedman, G. Tregear, R. Fernley, and A.E. Clarke, 1989. *Plant Cell* **1**: 483–491.

Boyes, D.C., C.H. Chen, T. Tantikanjana, J.J. Esch, and J.B. Nasrallah, 1991. *Genetics* **127**: 221–228.

Charlesworth, D. and B. Charlesworth, 1979. *Heredity* **43**: 41–55.

Clark, A.G. and T.-H. Kao, 1991. *Proc. Natl. Acad. Sci. U.S.A.* **88**: 9823–9827.

Clark, K.R., J.J. Okuley, P.D. Collins, and T.L. Sims, 1990. *Plant Cell* **2**: 815–826.

Coleman C.E. and T.-H. Kao, 1992. *Plant Mol. Biol.* **18**: 725–737.

Crow, J.F. and M. Kimura, 1970. *Introduction to Population Genetics Theory.* Harper and Row, New York

East, E.M. and A.J. Mangelsdorf, 1925. *Proc. Natl. Acad. Sci. U.S.A.* **11**: 166–171.

Ebert, P.R., M.A. Anderson, R. Bernatzky, M. Altschuler, and A.E. Clarke, 1989. *Cell* **56**: 255–262.

Emerson, S., 1938. *Genetics* **23**: 201–202.

Emerson, S., 1939. *Genetics* **24**: 535–537.

Ewens, W.J., 1964. *Genetics* **50**: 1433–1438.

Ewens, W.J., 1969. *Population Genetics*, pp. 71–76. Methuen, London.

Ewens, W.J. and P.M. Ewens, 1966. *Heredity* **21**: 371–378.

Fisher, R.A., 1958. *The Genetical Theory of Natural Selection*, pp. 104–110. Dover, New York

Fisher, R.A., 1961. *J. Theor. Biol.* **1**: 411–414.

Gillespie, J.H., 1989. *Am. Nat.* **134**: 638–658.

Graur, D., 1985. *J. Mol. Evol.* **22**: 53–62.

Griffiths, R.C. 1980. *Theor. Popul. Biol.* **17**: 37–50.

Gyllensten, U.B., M. Sundvall, and H.A. Erlich, 1991. *Proc. Natl. Acad. Sci. U.S.A.* **88**: 3686–3690.

Haring, V., J.E. Gray, B.A. McClure, M.A., Anderson, and A.E. Clarke, 1990. *Science* **250**: 937–941.

Hudson, R.R., 1983. *Theor. Popul. Biol.* **23**: 183–201.

Hudson, R.R., 1985. *Genetics* **109**: 611–631.

Ioerger, T.R., A.G. Clark, and T.-H. Kao, 1990. *Proc. Natl. Acad. Sci. U.S.A.* **87**: 9732–9735.

Jost, W., H. Bak, K. Glund, P. Terpstra, and J.J. Beintema, 1991. *Eur. J. Biochem.* **198**: 1–6.

Karlin, S., 1975. *Theor. Popul. Biol.* **7**: 364–398.

Kaufmann, H., F. Salamini, and R.D. Thompson, 1991. *Mol. Gen. Genet.* **226**: 457–466.

Kheyr-Pour, A., S.B. Bintrim, T.R. Ioerger, R. Remy, S.A. Hammond, and T.-H. Kao, 1990. *Sex. Plant. Reprod.* **3**: 88–97.

Kimura, M., 1965. *Ann. Rep. Nat. Inst. Genet. Jpn.* **16**: 86–88.

Kimura, M. and J.F. Crow, 1964. *Genetics* **49**: 725–738.

Kingman, J.F.C., 1982. *J. Appl. Prob.* **19A**: 27–43.

Lewis, D., 1948. *Heredity* **2**: 219–236.

Lewis, D., 1949. *Heredity* **3**: 339–355.

Lewis, D., 1951. *Heredity* **5**: 399–414.

Lewis, D., 1962. *J. Theror. Biol.* **2**: 69–71.

Li, W.-H., C.-I. Wu, and C.-C. Luo, 1985. *Mol. Biol. Evol.* **2**: 150–174.

Lipman, D.J., S.F. Altschul, and J.D. Kececioglu, 1989. *Proc. Natl. Acad. Sci. U.S.A.* **86**: 4412–4415.

Maruyama, T. and M. Nei, 1981. *Genetics* **98**: 441–459.

Mayo, A., 1966. *Biometrics* **22**: 111–120.

McClure, B.A., V. Haring, P.R. Ebert, M.A. Anderson, R.J. Simpson, F. Sakiyama, and A.E. Clarke, 1989. *Nature* **342**: 955–957.

McClure, B., J.E. Gray, M.A. Anderson, and A.E. Clarke, 1990. *Nature* **347**: 757-760.

Moran, P.A.P., 1962. *The Statistical Processes of Evolutionary Theory.* Clarendon Press, Oxford.

Nagylaki, T., 1975. *Genetics* **79**: 545-550.

Nasrallah, J.B. and M.E. Nasrallah, 1989. *Annu. Rev. Genet.* **23**: 121-139.

Nasrallah, J.B., T.-H. Kao, M.L. Goldberg, and M.E. Nasrallah, 1985. *Nature* **318**: 263-267.

Nasrallah, J.B., T.-H. Kao, C.-H. Chen, M.L. Goldberg, and M.E. Nasrallah, 1987. *Nature* **326**: 617-618.

Nei, M. and T. Gojobori, 1986. *Mol. Biol. Evol.* **3**: 418-426.

de Nettancourt, D., 1977. *Incompatibility in Angiosperms.* Springer-Verlag, Berlin, Heidelberg, New York.

Saitou, N. and M. Nei, 1987. *Mol. Biol. Evol.* **4**: 406-425.

Sawyer, S., 1989. *Mol. Biol. Evol.* **6**: 526-538.

She, J.X., S.A. Boehme, T.W. Wang, F. Bonhomme, and E.K. Wakeland, 1991. *Proc. Natl. Acad. Sci. U.S.A.* **88**: 453-457.

Shields, D.C., P.M. Sharp, D.G. Higgins, and F. Wright, 1988. *Mol. Biol. Evol.* **5**: 704-716.

Singh, A., Y. Ai, and T.-H. Kao, 1991. *Plant Phys.* **96**: 61-68.

Stephens, J.C., 1985. *Mol. Biol. Evol.* **2**: 539-556.

Tajima, F., 1989. *Genetics* **123**: 597-601.

Takahata, N., 1990a. In *Population Biology of Genes and Molecules*, edited by N. Takahata and J.F. Crow, pp. 267-285. Baifukan, Tokyo.

Takahata, N., 1990b. *Proc. Natl. Acad. Sci. U.S.A.* **87**: 2419-2423.

Takahata, N., 1991. In *New Aspects of the Genetics of Molecular Evolution*, edited by M. Kimura and N. Takahata, pp. 27-47. Springer-Verlag, Berlin.

Takahata, N. and M. Nei, 1990. *Genetics* **124**: 967-978.

Tavaré, S., 1984. *Theor. Popul. Biol.* **26**: 119-164.

Tsai, D.-S., L.C. Post, K.M. Kreiling, A.G. Clark, S.W. Schaeffer, and T.-H. Kao, 1991. *Plant. Mol. Biol.* (submitted).

Uyenoyama, M., 1988a. *Am. Nat.* **131**: 700-722.

Uyenoyama, M., 1988b. *Theor. Popul. Biol.* **34**: 47-91.

Uyenoyama, M., 1988c. *Theor. Popul. Biol.* **34**: 347-376.

Uyenoyama, M., 1991. *Genetics* **128**: 453-469.

Watterson, G.A., 1984. *Theor. Popul. Biol.* **26**: 387-407.

Wolfe, K.H., P.M. Sharp, and W.-H. Li, 1989. *J. Mol. Evol.* **29**: 208-211.

Wright, S., 1938. *Proc. Natl. Acad. Sci. U.S.A.* **24**: 253-259.

Wright, S., 1939. *Genetics* **24**: 538-552.

Wright, S., 1960. *Biometrics* **16**: 61-85.

Wright, S., 1964. *Evolution* **18**: 609-619.

Wright, S., 1969. *The Theory of Gene Frequencies*, pp. 402-416. University of Chicago Press, Chicago.

Xu, B., P. Grun, A. Kheyr-Pour, and T.-H. Kao, 1990a. *Sex. Plant. Reprod.* **3**: 54-60.

Xu, B., J.H. Mu, D.L. Nevins, P. Grun, and T.-H. Kao, 1990b. *Mol. Gen. Genet.* **224**: 341-346.

Yokoyama, S. and M. Nei, 1979. *Genetics* **91**: 609-626.

Yokoyama, S. and L.E. Hetherington, 1982. *Heredity* **48**: 299-303.

Coevolution of Immunogenic Proteins of *Plasmodium falciparum* and the Host's Immune System

AUSTIN L. HUGHES

Department of Biology and Institute of Molecular Evolutionary Genetics,
The Pennsylvania State University, University Park, PA 16802, U.S.A.

The malaria parasite *Plasmodium falciparum* is one of the most geographically widespread eukaryotic parasites of humans (Molineaux, 1988). Even in the age of AIDS, *falciparum* malaria remains the major pathogenic killer of humans in the tropics (Kreier and Baker, 1987). In achieving this high level of success as a parasite, *P. falciparum* must have evolved adaptations enabling it to overcome the defenses of its host's immune system. Here, I examine recent findings at the molecular level regarding such adaptations, with emphasis on coevolution between *P. falciparum* proteins and the major histocompatibility complex (MHC) of the vertebrate host.

Coevolution between *P. falciparum* and its host's immune system is of interest not only as an illustration of adaptive evolution at the molecular level but also because it can provide clues regarding the history of the association between *P. falciparum* and humans. It has frequently been speculated that *P. falciparum* has recently switched to humans from another vertebrate host (Coatney *et al.*, 1971). Here, I consider to what extent the molecular data are consistent with this hypothesis. Klein (1991) has recently proposed that MHC alleles have co-evolved with specific alleles of protozoan parasites, many of which seem to have a clonal population structure (Tibayrenc *et al.*, 1990). Analysis of DNA sequence data from genes encoding immunogenic proteins can shed light on the applicability of this model of evolution to *P. falciparum*.

BIOLOGY OF *P. FALCIPARUM*

The genus *Plasmodium* (Apicomplexa: Sporozoea: Eucoccidiida: Haemosporina: Plasmodiidae) contains parasites whose life cycle involves a mosquito (Diptera: Culicidae) and a vertebrate host (Kreier and Baker, 1987). There are four species infecting humans: *P. vivax*, *P. malariae*, *P. ovale*, and *P. falciparum*. These and other *Plasmodium* infecting primates are transmitted by mosquitos of the genus *Anopheles*. Infection of the vertebrate host begins with injection of sporozoites from the mosquito's salivary glands during a blood meal. The sporozoites, which are haploid, enter the liver parenchyma cells; there the sporozoites (now called trophozoites) divide mitotically and eventually cause rupture of the host cell and release of merozoites. The merozoites enter red blood cells, and there are a number of cycles of mitotic reproduction, release of merozoites, and invasion of further red blood cells. Some merozoites in red blood cells develop into male microgametes or female macrogametes. If these are taken up by a mosquito in a blood meal, fertilization may take place in the midgut of the mosquito. The zygote enters the lining of the gut, where reductional division takes place, giving rise to sporozoites, which migrate throughout the mosquito (Garnham, 1988).

This life cycle, involving zygote formation in the mosquito, is so far the only one known for *Plasmodium*. Recently, population genetic studies on parasitic protozoa have suggested that several of these species may have clonal population structures since segregation and recombination appear to be rare in natural populations of these species (Tibayrenc *et al.*, 1990). According to Tibayrenc *et al.* (1991), "the available evidence is contradictory" in the case of *P. falciparum*; but "the possibility of an unknown uniparental cycle cannot be discarded."

Evolutionary relationships within the genus *Plasmodium* were previously completely unknown, but are now beginning to be resolved by molecular evidence. McCutchan *et al.* (1984) showed that *P. falciparum* is more similar to rodent and avian malarias with respect to genomic $G+C$ content than it is to *P. vivax* and several *Plasmodium* species parasitic on nonhuman primates. Thus, they suggested that *P. falciparum* may not be closely related to other primate malarias and that host and parasite phylogenies might not be parallel in the case of this genus. On the basis of small subunit (SSU) rRNA genes, Waters, Higgins, and McCutchan (1991) produced a phylogenetic tree grouping *P. falciparum* with the avian parasites *P. gallinaceum* and *P. lophurae* (Fig. 1). In this phylogenetic tree, the cluster of *P. falciparum* and avian malarias was next most

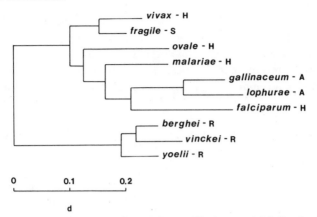

Fig. 1. Phylogenetic tree (redrawn from Waters, Higgins, and McCutchan) based on number of nucleotide substitutions per site (*d*) in a species-specific region of SSU rRNA gene. The tree was constructed by the neighbor-joining method (Saitou and Nei, 1987).

closely related to *P. malariae* and *P. ovale*. Three species of rodent malarias were separated by a long internal branch from the primate and avian malarias, and thus I have placed the root of the tree on the branch separating the rodent malarias from other species considered (Fig. 1).

Waters, Higgins, and McCutchan suggest that their phylogeny provides evidence for a lateral transfer of *P. falciparum* from an avian to a human host. While recognizing that there is a "formal possibility that the transfer occurred in the opposite direction," they argue that this is unlikely. However, the hypothesis that the ancestor of the closely related avian malarias *P. gallinaceum* and *P. lophurae* was transferred to birds from a primate host seems more parsimonious if their phylogenetic tree is correct. Given the tree's topology (Fig. 1), if we suppose that *P. falciparum* was transferred to humans from a bird host, we must also assume that a short time prior to this the ancestor of the same avian malaria was transferred to birds from a primate host. The tree could only be consistent with Waters, Higgins, and McCutchan's hypothesis if it were re-rooted so that the *P. falciparum*/*P. gallinaceum*/*P. lophurae* cluster was an outgroup. But if this were done, the branch leading to the rodent malarias would be implausibly long.

It has often been surmised that *P. falciparum* has been a human parasite for only a short time, possibly only since the development of agriculture in West Africa about 5,000 years ago (Livingstone, 1958). The following lines of reasoning have been used in support of this hypothesis:

(1) Traditionally parasitologists have held that virulence characterizes a

recently acquired parasite (Cameron, 1956). The fact that *P. falciparum* is the most virulent of all human malarias has thus been taken as evidence that this parasite has been acquired by humans relatively recently (Coatney *et al.*, 1971). Although some recent mathematical models have supported the view that parasites evolve in the direction of commensalism, May (1985) emphasizes that this conclusion is dependent on the assumptions of the models and need not be true in general.

(2) Data on frequency of the sickle-cell hemoglobin gene in West Africa are consistent with the idea that the increase in frequency of this gene as a result of selection favoring sickle-cell heterozygotes (arising from *P. falciparum*) has occurred over about the past 6,000 years. This is approximately the time since the arrival of agriculture in the region (Livingstone, 1958). However, this need not imply that *P. falciparum* was not present previously in human populations, although the frequency of infection may indeed have increased in agricultural populations in comparison with hunter-gatherer populations. Further, the sickle-cell mutation may not have occurred previously; or it may have occurred previously but may then have been eliminated by chance. Thus at present there is no compelling evidence for the view that the association between *P. falciparum* and its human host is particularly recent.

Because of the interest in developing a vaccine against *P. falciparum*, there have been a number of recent studies of *P. falciparum* proteins evoking an immune response in humans. The most intensively studied of these has been the circumsporozoite (CS) protein. The CS protein covers the surface of the mature sporozoite (Yoshida *et al.*, 1980), accounting in one species for 10–20% of the protein synthesized by the sporozoite (Cochrane *et al.*, 1982). Several merozoite surface proteins have also recently been studied at the molecular level, including the merozoite surface antigen-1 (MSA-1). The CS protein and MSA-1 are each encoded by a single polymorphic locus. The fact that the sporozoite and merozoite have different surface antigens represents a simple form of an antigen-shifting strategy. Such strategies are characteristic of protozoan parasites of vertebrates and are generally considered as means to evade immune recognition. The greatest elaboration of antigen-shifting is seen in the variant surface glycoproteins of trypanosomes, in which genetic re-arrangements serve to generate diversity over the course of an infection (Cross, 1990). So far, there is no evidence of any similarly elaborate mechanism of host-evasion in any *Plasmodium* species.

INTERACTIONS WITH THE VERTEBRATE IMMUNE SYSTEM

A successful parasite of vertebrates must evade two separate recognition systems: (1) the immunoglobulin system; and (2) the MHC and T-cell receptor (TCR) system. The Ig system involves recognition and immobilization of extracellular foreign protein and nonprotein molecules. The MHC-TCR system involves recognition of short peptides derived from foreign proteins that have been broken down within host cells; the foreign peptide is bound to the MHC molecule and transported to the cell surface, where this complex is recognized by the TCR. There are two functional divisions in the MHC: class I and class II. Class I MHC molecules are expressed on the surface of all nucleated somatic cells, and the foreign peptides they present are derived from pathogens causing an intracellular infection. The class II MHC molecules are expressed on antigen-presenting cells of the immune system, and the foreign peptides they present are derived from foreign proteins that have been taken up by these cells (Klein, 1986).

These two recognition systems differ in a fundamental way in their ability to recognize a variety of foreign antigens. Immunoglobulins are generated by recombination of DNA segments and by additional somatic mutation; as a result, they have an extraordinary variety and virtually limitless ability to recognize foreign antigens. The TCR also are generated by recombination of DNA segments (though without somatic mutation) and thus also have the potential for enormous variety. On the other hand, the MHC of a given individual includes a relatively limited number of possible peptide-binding domains. In mammals, there are 1-3 functional class I MHC loci, each of which is generally polymorphic; thus 2-6 types of class I MHC molecules are expressed by a given mammal. Similarly, a given mammal may express 2-8 class II MHC molecules. The evolutionary factors responsible for this limitation on MHC diversity are still speculative (Howard, 1987); but it is clear that, in order to provide adequate immune protection, this limited number of molecules must be able to bind at least one peptide from each parasite the individual is likely to encounter. Note, however, that the antigen-binding by MHC molecules lacks the specificity of binding by immunoglobulins. Rather, each MHC molecule binds a series of peptides sharing common structural features (Sette *et al.*, 1988).

Here, I consider current ideas on the way that immunogenic proteins of *Plasmodium* interact with these two recognition systems.

1. Amino Acid Repeats and the Smoke-Screen Hypothesis

One of the most striking characteristics of immunogenic proteins of parasitic protozoa is the frequent occurrence of domains including numerous repeats of a short amino acid motif (Kemp *et al.*, 1987). In the CS protein of *P. falciparum*, for example, there is a region consisting of 40–48 repeats (depending on the allele) of a four amino acid sequence (usually Asn-Ala-Asn-Pro, with occasional variants). This repeat region is exposed on the surface of the sporozoite and is the predominant target of antibodies (Ballou *et al.*, 1985), yet it is known that immunization against this repeat region alone does not confer full protection against infection.

These facts regarding the repeat region of the CS protein have led to the formulation of a hypothesis to explain the evolution of such regions in immunogenic proteins of parasitic protozoa called the "smoke-screen hypothesis" (Kemp *et al.*, 1987). Simple amino acid polymers are known to elicit a rather ineffective T cell-independent antibody response, and it is argued that the function of repeat regions is to elicit such a response (Enea and Arnot, 1988). Presumably a T cell-dependent response against non-repeat regions would be more effective (Vergara *et al.*, 1986), but induction of a strong T cell-independent response to the repeat region appears to prevent development of a T cell-dependent response (Enea and Arnot, 1988). Furthermore, antibodies against repeat regions of one protein may cross-react with those of other proteins and thus appear to lack specificity (Kemp *et al.*, 1987).

Different CS protein alleles in *P. falciparum* have similar repeat region sequences, although they may differ slightly in number of repeats (de la Cruz *et al.*, 1987). In the case of *P. cynomolgi*, a parasite of monkeys mainly in the genus *Macaca*, the amino acid sequence of repeats differs markedly among alleles (Galinski *et al.*, 1987). There is some evidence in the pattern of nucleotide substitution among these alleles that natural selection has favored diversification of the repeat region among *P. cynomolgi* CS alleles (Hughes, 1991). However, it is still unknown why these two species differ so strikingly in the repeat region.

2. The Role of the MHC

Both class I and class II MHC gene families in humans contain certain highly polymorphic loci. Many hypotheses have been proposed to explain this high polymorphism, but only recently has it been possible to rule out several of these (such as the idea that the mutation rate in the MHC is unusually high and the idea that inter-locus "gene conversion" alone can account for MHC polymorphism). Comparison of DNA sequences shows that the rate of nonsynonymous nucleotide substitution per site (d_N) exceeds that of synonymous substitution per site (d_S) in the codons encoding the putative antigen-binding portions of the molecule (Hughes and Nei, 1988, 1989; Hughes *et al.*, 1990). In other regions of the same genes, $d_S > d_N$, as is true of most genes because purifying selection eliminates most nonsynonymous mutations. This result indicates that positive selection favoring diversity at the amino acid level is acting on precisely those regions of the molecule which bind foreign peptides. This in turn suggests that MHC polymorphism is maintained by some form of balancing selection relating to disease resistance.

The most biologically realistic model of such selection is the model of overdominant selection (heterozygote advantage). It is based on the fact that MHC alleles differ in their ability to present different peptides (Doherty and Zinkernagel, 1975). Thus, in a population exposed to an array of pathogens, a heterozygote at all or most MHC loci would have an advantage in being able to bind a wide array of peptide types and thus being protected against most pathogens. Evidence of association between presence of a particular MHC allele or haplotype and resistance to specific diseases (including *falciparum* malaria; Hill *et al.*, 1991) is consistent with this hypothesis.

One characteristic of the MHC is that in humans, mice, and some other species, MHC polymorphisms have been found to be very ancient, predating the origin of the species in question. For example, certain human MHC alleles show evidence of a closer relationship to certain alleles from the orthologous locus in the chimpanzee than they show to other human alleles from the same locus (Mayer *et al.*, 1988). The existence of such long-lasting polymorphism is itself evidence of balancing selection, since neutral polymorphisms are unlikely to be maintained for such long periods (Takahata and Nei, 1990).

It is frequently argued that balancing selection at MHC loci must be frequency-dependent rather than overdominant. Mathematically, one model of frequency-dependent selection (rare allele advantage) is very similar to overdominant selection, and either model could explain both the large number of alleles at MHC loci and their long persistence (Takahata and Nei, 1990). However, Takahata and Nei (1990) question the biological realism of the model of rare allele advantage. Hill *et al.* (1991) argue that their data on the association between MHC genotype and resistance to *falciparum* malaria support the hypothesis of frequency-dependent selection. In fact, since the resistant alleles and haplotypes are actually quite common, their results do not support the model of rare allele advantage. A true test of the overdominance model requires data on resistance to at least two different pathogens. In the case of *P. falciparum*, these might conceivably be strains bearing different alleles at the

CS protein locus, or other loci involved in immune resistance. So far, however, no such data are available for natural populations of any vertebrate species.

In highly conserved proteins, amino acid replacements tend to be conservative; that is, when an amino acid is replaced, it tends to be replaced by a chemically similar amino acid (Miyata *et al.*, 1979). In MHC binding regions, on the other hand, where selection actively favors diversity at the amino acid level, it might be expected that radical (nonconservative) amino acid replacements might be favored. In fact, Monos *et al.* (1984) noted that class I MHC alleles often have charge differences among them. Hughes *et al.* (1990) developed a method to test whether nonsynonymous nucleotide differences causing a difference in residue charge or some other property of interest occur more frequently than expected under random substitution. This method depends on computing the proportion of conservative nonsynonymous differences (p_{NC}) and the proportion of radical nonsynonymous differences (p_{NR}). When $p_{NR} > p_{NC}$, radical nonsynonymous differences are found more frequently than expected by chance.

When this test was applied with respect to residue charge in the antigen binding cleft of the class I MHC molecules of humans and mice, p_{NR} was found to exceed p_{NC} significantly. This suggests that natural selection has acted to diversify charge profile (the pattern of charged residues) among class I binding clefts (Hughes *et al.*, 1990). When a similar analysis was applied to class II MHC alleles, there was no significant tendency to favor charge change in putative binding regions; in this case, binding regions were found to have numerous amino acid differences involving both charge changes and other changes (Hughes, unpublished data).

The function of the class I MHC molecule is to mark a cell infected by an intracellular pathogen so that cytotoxic (CD8+) T cells can recognize the cell (by binding with the complex of foreign peptide and MHC molecule) and kill it. On the other hand, the complex of class II MHC molecule and foreign peptide is recognized by a helper (CD4+) T cell. The helper T cell then stimulates an appropriate immune response, including proliferation of B cells producing immunoglobulins specific for some antigen from the same pathogen from which the MHC-bound peptide was derived.

A number of recent studies have focused on the role of the two MHC classes in response to infection with *Plasmodium*. Peptides derived from the CS protein of *P. falciparum* have been shown to cause proliferation *in vitro* of T cells derived from individuals previously exposed to this parasite (Hoffman *et al.*, 1989; DeGroot *et al.*, 1989). Such T cell proliferation is triggered by CD4+ T cell recognition of a peptide bound by the class II MHC. The most effective peptides in this assay are identified as the putative T cell epitopes (TCE) of the CS protein. A role for the class I MHC in resistance to *Plasmodium* was indicated by experiments in which mice whose CD8+ T cells were depleted were unable to resist *P. yoelii* infection, whereas depletion of CD4+ T cells had no such detrimental effect (Weiss *et al.*, 1988). Presumably, cytotoxic T cells play a role during the infection of liver cells, since red blood cells do not express MHC molecules. In addition, cytotoxic T cells from mice infected with *P. falciparum* recognized a peptide overlapping the known CD4+ TCE (Kumar *et al.*, 1988).

POSITIVE SELECTION ON IMMUNOGENIC PROTEINS

In order to test whether selection to evade recognition by the host MHC has acted on the CS protein of *P. falciparum*, rates of synonymous and nonsynonymous nucleotide substitution were estimated in pairwise comparisons among CS alleles; rates were computed separately for the TCE and other gene regions (Table I). Among *P. falciparum* CS protein alleles for which complete sequence is available, there are no synonymous differences in non-repeat regions; but the rate of nonsynonymous substitution is greatly enhanced in the TCE over other regions (Table I). When the same regions of alleles from *P. cynomolgi* were compared, no particular elevation of the nonsynonymous rate was found in the regions corresponding to the *P. falciparum* TCE. The results thus suggest that positive selection to evade host recognition acts on the TCE of *P. falciparum* CS proteins. If there is similar selection in the case of *P.*

TABLE I

Mean Numbers of Nucleotide Substitutions per 100 Synonymous Sites (d_S) and per 100 Nonsynonymous Sites (d_N) and Mean Percent Conservative (p_{NC}) and Radical (p_{NR} Nonsynonymous Difference, with Respect to Residue Charge, in Comparison of CS Protein Alleles of *P. falciparum* and *P. cynomolgi*

Comparison	5'NR ($n=65$)		3'NR (excluding TCE) ($n=42$)		TCE ($n=36$)	
	d_S	d_N	d_S	d_N	d_S	d_N
P. falciparum	0.0 ± 0.0	0.6 ± 0.5	0.0 ± 0.0	0.8 ± 0.6	0.0 ± 0.0	5.7 ± 1.7***
P. cynomolgi	10.9 ± 3.7	5.2 ± 1.3	5.5 ± 3.2	2.3 ± 1.0	1.7 ± 1.7	3.4 ± 1.4
	p_{NC}	p_{NR}	p_{NC}	p_{NR}	p_{NC}	p_{NR}
P. falciparum	0.8 ± 0.8	0.5 ± 0.5	1.3 ± 1.0	0.0 ± 0.0	0.9 ± 0.9	9.7 ± 3.1**
P. cynomolgi	4.8 ± 1.7	5.1 ± 1.8	3.1 ± 1.5	1.0 ± 1.0	6.6 ± 2.7	0.0 ± 0.0*

For details on regions analyzed, see Hughes (1991). d_S and d_N were estimated by Nei and Gojobori's (1986) method. *$p<0.05$; **$p<0.01$; ***$p<0.001$.

cynomolgi it does not act on exactly the same region. Indeed, examination of *P. cynomolgi* sequences revealed a region where the rate of nonsynonymous substitution significantly exceeds the synonymous rate (Hughes, 1991).

In addition, in the TCE of *P. falciparum* CS protein alleles, p_{NR} with respect to residue charge significantly exceeds p_{NC} (Table I). Thus selection acts to diversify the profile of residue charges in this region. Given the importance of charge differences in the binding cleft among MHC alleles (Hughes *et al.*, 1990), this finding supports the hypothesis that selection favoring diversity of the TCE is based on evading binding by the MHC (Hughes, 1991).

Analysis of MSA-1 alleles from *P. falciparum* reveals a complex history of recombination events (Hughes, 1992). These include both nonreciprocal recombination events that have homogenized certain regions among known alleles and reciprocal recombination events that have exchanged certain regions among alleles. The MSA-1 alleles can be categorized into two groups (Group 1 and Group 2) based on their similarity in the most divergent regions of the gene. For the 5' end of the gene, sequences are available for three alleles in each group. Table II shows d_S and d_N within and between these two groups in seven 5' regions of the gene. For definition of the regions, see Hughes (1992).

The regions differ strikingly in degree of difference between the two groups at both synonymous and nonsynonymous sites. In Region 3, there has been a

TABLE II
Mean Numbers of Synonymous (d_S) and Nonsynonymous (d_N) Substitutions per 100 Sites ($\pm$S.E.) in Comparisons of 5' Regions of MSA-1 Gene Alleles

Region (No. codons)	Group 1 (3)		Group 2 (3)		Group 1 *vs.* Group 2 (9)	
	d_S	d_N	d_S	d_N	d_S	d_N
1 (18)	0.0±0.0	0.0±0.0	0.0±0.0	0.0± 0.0	0.0±0.0[c]	0.0±0.0[c]
2 (47)	1.5±1.9	0.8±0.4	6.1±2.1[b]	3.7± 0.9[b,e]	7.5±2.1[c]	2.8±0.6[*c]
3 (49)	12.8±5.1[c]	50.9±8.9[***c]	30.4±8.6[c]	83.5±10.7[***c,d]	10.8±3.9[c]	56.5±7.2[***]
4 (54)	0.0±0.0	0.0±0.0	0.0±0.0	0.5±0.5	0.0±0.0[c]	6.5±2.1[**c]
5 (64)	0.0±0.0	0.8±0.6	2.8±5.1[a]	21.6±3.3[c,f]	8.5±3.0[c]	16.5±2.7[*c]
6 (62)	0.0±0.0	0.1±0.1	0.3±0.2	0.8±0.2[e]	63.6±6.0	46.0±2.2[**]
7 (58)	0.0±0.0	0.0±0.0	0.0±0.0	0.0±0.0	14.2±7.3[c]	10.5±2.9[c]

Figures in parentheses are numbers of comparisons. Tests of significance of the difference between d_S and d_N: [*]$p<0.05$; [**]$p<0.01$; [***]$p<0.001$. Tests of significance of the difference between a value of d_S or d_N and the corresponding value for Region 6: [a]$p<0.05$; [b]$p<0.01$; [c]$p<0.001$. Tests of significance of the difference between a value of d_S or d_N in comparisons among group II alleles and the corresponding value among group I alleles: [d]$p<0.05$; [e]$p<0.01$; [f]$p<0.001$. Standard errors of mean d_S and d_N are estimated by Nei and Jin's (1989) method. From Hughes (1992).

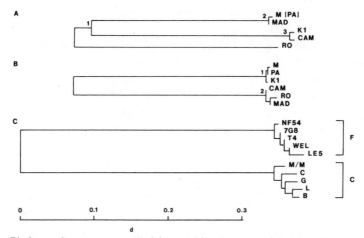

Fig. 2. Phylogenetic trees constructed by neighbor joining method based on number of nucleotide substitutions per site (d). (A) Tree for Region 3 of MSA-1 alleles. The lengths of branches 1-2 (0.252 ± 0.063) and 1-3 (0.266 ± 0.065) are significantly different from zero at the 0.1% level (method of Li, 1989). (B) Tree for Region 6 of MSA-1 alleles. The length of the branch 1-2 (0.522 ± 0.023) is significantly different from zero at the 0.1% level. (C) Tree for non-repeat region (143 aligned codons) of CS protein gene alleles for two *Plasmodium* species: *P. falciparum* (F) and *P. cynomolgi* (C). From Hughes (1992).

TABLE III

Mean Numbers of Synonymous (d_S: below diagonal) and Nonsynonymous (d_N: above diagonal) Nucleotide Substitutions per 100 Sites ($\pm$S.E.) in Pairwise Comparisons among Region 3 of MSA-1 Gene Alleles

	M	PA	K12	MAD	CAM	RO
M		0.0 ± 0.0	$76.4 \pm 13.4^{***}$	3.9 ± 2.0^a	$76.4 \pm 13.4^{a***}$	$89.3 \pm 15.9^{***}$
PA	0.0 ± 0.0		$76.4 \pm 13.4^{***}$	3.9 ± 2.0^a	$76.4 \pm 13.4^{a***}$	$89.3 \pm 15.9^{***}$
K1	19.2 ± 7.6	19.2 ± 7.6		$76.4 \pm 13.4^{***}$	1.9 ± 1.3	90.7 ± 16.0
MAD	4.8 ± 3.4	4.8 ± 3.4	13.0 ± 6.1		$76.4 \pm 13.4^{a***}$	$83.3 \pm 14.7^{***}$
CAM	19.2 ± 7.6	19.2 ± 7.6	0.0 ± 0.0	13.1 ± 6.1		90.7 ± 16.0
RO	14.5 ± 6.5	14.5 ± 6.5	6.9 ± 4.4	8.8 ± 4.9	6.9 ± 4.4	

Tests of significance of the difference between d_S and d_N: *$p < 0.05$; **$p < 0.01$; ***$p < 0.001$. Tests of significance of the difference between a value of d_N and the value of d_N for the comparison with K1 in the same column: [a]$p < 0.001$. From Hughes (1992).

reciprocal exchange so that the Group 2 allele MAD resembles Group 1, and the Group 1 allele K1 resembles Group 2 (Table II and Fig. 2). Likewise in Region 3, the Group 2 allele RO resembles neither group, having received this region from a donor apparently belonging to neither of the two major groups (Table II and Fig. 2).

In three regions of the MSA-1 gene, d_N significantly exceeds d_S in the comparison between the two groups of alleles. This is evidence of positive selection favoring diversity between the two groups. The three regions are Regions 3, 4, and 5 (Tables II and III). Region 3 has a repeat structure. In addition to a high rate of nonsynonymous substitution in aligned codons, alleles differ in length in this region due to insertion or deletion events (Fig. 3).

A

```
M     NEGTSGTAVTTSTPGSKGSVASGGSGGS---------------------------VASGGSVASGGS
PA    NEGTSGTAVTTSTPGSKGSVASGGSGGS---------------------------VASGGSVASGGS
K1    NEEEITTKGASAQSGTSGT----------------------------------------SGTSGPSGPSGT
MAD   NEGTSGTAVTTSTPGSSGSVT---SGGS-------------------------------VASVASVASGG-
CAM   NEEEITTKGASAQSGTSGTSGTSGTSGTSGTSGTSGTSAQSGTSGTSGTSAQSGTSGTSAQSGTSGTSGTSGT
RO    KDGANTQVVAKPADAVSTQ-------------------------------------------SAKNPPGAT


M     VASGGSVASGGSGNSRRTNPSDNS
PA    VASGGSVASGGSGNSRRTNPSDNS
K1    SPSSRSNTLPRSNTSSGASPPADA
MAD   --SGGSVASGGSGNSRRTNPSDNS
CAM   SPSSRSNTLPRSNTSSGASPPADA
RO    VPSGTASTKGAIRSPGAANPSDDS
```

B

```
M     KTIENINELIEESKKTIDKNKNATKEEEKKKLYQAQYDLSIYNKQLE
      *  *       *     *    ** *              *  :       *
MAD   TTIANINELIEGSKKTIDQNKNADNEEGKKKLYQAQYNLFIYNKQLQ
```

Fig. 3. MSA-1 regions showing evidence of positive selection. (A) Region 3 (tripeptide repeat region). (B) Comparison of most variable portion of Region 4 from a group I allele (M) and a group II allele (MAD): * = amino acid replacement involving a residue charge change; : = amino acid replacement not involving a charge change.

TABLE IV

Mean Percent Conservative (p_{NC}) and Radical (p_{NR}) Nonsynonymous Nucleotide Substitutions ($\pm$S.E.) with Respect to Amino Acid Residue Charge in 5' Regions of MSA-1 Gene Alleles

Region	Group 1 (3)		Group 2 (3)		Group 1 vs. Group 2 (9)	
	p_{NC}	p_{NR}	p_{NC}	p_{NR}	p_{NC}	p_{NR}
3	32.2±4.0	31.5±5.6	54.9±5.5	41.8±7.3	37.2±3.8	29.5±5.0
4	0.0±0.0	0.0±0.0	0.0±0.0	0.0±0.0	1.7±1.7	9.8±3.4*
5	0.8±0.8	0.9±0.9	18.7±3.6	18.4±3.7	14.3±2.9	13.7±2.8

Figures in parentheses are numbers of comparisons. Test of significance of difference between p_{NC} and p_{NR}: *$p < 0.05$. From Hughes (1992).

Region 4 is a region of comparatively high sequence similarity between the two groups of alleles (Table III). In this region p_{NR} with respect to charge also significantly exceeds p_{NC} (Table IV). Figure 3 shows a variable portion of Region 4 in which a high preponderance of amino acid differences involve charge differences.

The repeating structure of Region 3 suggests it may have evolved as a "smoke-screen" to attract immunoglobulin response. Region 4, on the other hand, may include epitopes bound by the MHC and recognized by T cells. The reason for making such a supposition is that its characteristics are similar to those of the TCE of the CS protein. So far, two peptides from MSA-1 have been shown to be recognized by T cells (Crisanti *et al.*, 1988); these are from nonpolymorphic portions of Regions 2 and 4.

AGE OF POLYMORPHISMS IN *P. FALCIPARUM*

Because the MHC is characterized by long-lasting polymorphisms and because both the CS protein and MSA-1 appear to have evolved polymorphisms in response to selection for evasion of MHC binding, it would be of interest to know how the persistence time of CS protein and MSA-1 alleles compares to that of MHC alleles. Unfortunately, since there is no fossil record for *Plasmodium*, inferences regarding the age of alleles must be made rather indirectly; thus any conclusions must be more tentative than in the case of vertebrates.

In the case of the CS protein, I used d_N in non-repeat regions exclusive of the TCE to obtain information regarding the age of alleles in *P. falciparum* (Table V). If the rodent malarias *P. yoelii* and *P. berghei* are assumed to have diverged from *P. falciparum* when rodents and primates diverged (about 80 million years ago; Li *et al.*, 1990), we would obtain an estimate of 1.9 million years for the divergence time of the two most divergent *P. falciparum* CS protein alleles. In this case, the rate of nonsynonymous substitution would be about 3×10^{-9} substitutions per site per year.

This estimate may be questioned since it depends on the assumption that the CS protein (excluding the *P. falciparum* TCE) is a good clock. Although the region of elevated nonsynonymous substitution (TCE) for *P. falciparum* has been excluded, other species may have analogous selected regions in other parts of the gene. Since these are not excluded, the divergence time among *P. falciparum* alleles may be underestimated. Furthermore, the fact that d_N between the CS protein of *P. falciparum* and that of *P. vivax* is actually higher than that between *P. falciparum* and the much more distantly related *P. yoelii*

TABLE V

Mean Numbers of Nonsynonymous Substitutions per 100 Sites ($\pm$S.E.) in Regions on Five *Plasmodium falciparum* CS Protein Alleles and between These Alleles and CS Protein Alleles from Four Other *Plasmodium* Species

	T4	WEL	LE5	7G8
NF54	1.3 ± 0.8	0.9 ± 0.6	0.9 ± 0.6	0.9 ± 0.6
T4		0.4 ± 0.4	1.3 ± 0.8	0.4 ± 0.4
WEL			0.9 ± 0.6	0.0 ± 0.0
LE5				0.9 ± 0.6
	cynomolgi	*vivax*	*yoelii*	*berghei*
falciparum	64.9 ± 7.6	65.7 ± 7.9	50.2 ± 6.2	57.3 ± 6.9
cynomolgi		15.4 ± 2.7	58.2 ± 6.9	54.4 ± 6.5
vivax			69.0 ± 8.3	59.4 ± 7.2
yoelii				22.2 ± 3.5

The *P. falciparum* TCE is excluded from analysis. Number of codons compared = 107. There are five alleles from *P. cynomolgi*, one each from the other three species.

and *P. berghei* (Fig. 1 and Table V) argues against a simple relation between d_N in this gene and divergence time. It may be that there are only a few functional constraints on the CS protein, but that over a long evolutionary time forward and backward replacements have occurred at many positions, making the estimate of d_N inaccurate in comparisons as distant as those between *P. falciparum* and the rodent malarias.

In any event, even if this estimate of the divergence time between *P. falciparum* alleles is 2–4 times too high, polymorphism at the CS protein locus is predicted to antedate considerably the origin of *Homo sapiens*. However, the CS polymorphism is not older than that at human MHC loci, which is at least 5 million years old (Mayer *et al.*, 1988).

In the case of MSA-1, alleles are mosaics of recently homogenized regions and much older regions. On the basis of substitutions between the two groups of alleles at both synonymous and nonsynonymous sites, Region 6 appears to be the most anciently diverged region (Table II). MSA-1 sequences are available for *P. yoelii* and another rodent malaria *P. chabaudi*. Similarity between these genes and those of *P. falciparum* is low, but Lewis (1989) has identified two conserved portions of Region 6. Table VI shows d_S and d_N in this region between the *P. yoelii* and *P. chabaudi* sequences and between each of them and the two groups of *P. falciparum* alleles. Assuming a divergence time of 80 million years between *P. falciparum* and the rodent malarias, comparison of d_N values provides an estimate of about 35 million years for the divergence in Region 6 between the two groups of *P. falciparum* MSA-1 alleles. This would correspond to a rate of 2.6×10^{-9} nonsynonymous substitutions per site per

TABLE VI

Comparisons of Numbers of Synonymous (d_S: above diagonal) and of Nonsynonymous (d_N: below diagonal) Substitutions per 100 Sites in a Conserved Region among MSA-1 Alleles from Two Rodent Malarias and the Two Major Groups of *P. falciparum* Alleles

	P. yoelii	*P. chabaudi*	*P. falciparum* Group 1	*P. falciparum* Group 2
P. yoelii	—	20.2 ± 4.6	76.7 ± 13.6	63.7 ± 11.2
P. chabaudi	5.3 ± 1.2	—	85.8 ± 15.4	75.1 ± 13.1
P. falciparum Group 1	44.1 ± 4.1	45.7 ± 4.2	—	68.1 ± 12.2
P. falciparum Group 2	36.8 ± 3.6	39.2 ± 3.7	18.1 ± 2.2	—

Number of codons compared = 179. From Hughes (1992).

year. Note that in this case, synonymous sites probably cannot be used to provide an estimate of divergence time; d_S estimates between *P. falciparum* and the rodent malarias are very high and thus probably not very accurate.

If the MSA-1 polymorphism has been maintained for 35 million years, it is older than most MHC polymorphisms. It is possible that primate *DQA1* polymorphisms have been maintained equally long or longer (Gyllensten and Ehrlich, 1989). Simulations suggest that balancing selection can maintain a polymorphism involving a small number of alleles for a longer time than a polymorphism with numerous alleles (Takahata and Nei, 1990). Thus it is of interest that the MSA-1 polymorphism in *P. falciparum*, which in Region 6 involves essentially only two allelic types, is one of the oldest known in any organism.

In the repeat region (Region 3), MSA-1 alleles are about as divergent at nonsynonymous sites as they are in Region 6. Here, however, since selection has favored diversification, nonsynonymous sites may have evolved more rapidly than in Region 6. Thus, the polymorphism in this region may not be as old as in Region 6. In fact, on the basis of the synonymous rate, the polymorphism in Region 3 may be not much more than 5 million years old. In Region 4, which may contain T cell epitopes, the divergence between Group 1 and Group 2 alleles must be far more recent. If we can assume that the rate of synonymous evolution is the same in Regions 4 and 6, we might estimate that the two groups of alleles diverged in Region 4 no more than about half a million years ago.

CONCLUSIONS

As mentioned, the SSU rRNA phylogeny of *Plasmodium* species (Fig. 1) does not support the hypothesis that *P. falciparum* has recently transferred to humans from some other host species. The fact that both the CS protein and MSA-1 loci have polymorphisms that antedate the origin of modern humans is also not easy to explain on the hypothesis that *P. falciparum* was recently transferred to *H. sapiens*. Nothing is so far known about how a *Plasmodium* parasite might be transferred from one vertebrate host to another. But it seems likely that this would be a rare event, and that only a small number of parasites would be involved. If so, such a transfer would involve a severe population bottleneck and accompanying severe reduction of polymorphism.

The fact that the evolutionary history of *P. falciparum* MSA-1 has involved repeated events of interallelic genetic exchange provides clear evidence of sexual reproduction. The data do not, however, rule out the possibility of an asexual mode of reproduction in this species. If recombinants at the MSA-1 locus have been selectively favored, they may be observed even if such events are relatively rare. In bacteria, loci involved in interaction with the host immune system show greater evidence of recombination (Smith *et al.*, 1990) than do housekeeping enzyme loci (Nelson *et al.*, 1991); the discrepancy can be explained by selection favoring recombinant types at the former loci.

Whatever the population structure of *P. falciparum*, the data on the CS protein and MSA-1 are not consistent with the hypothesis of a long-lasting coevolution between particular parasite antigens and particular MHC genes. In the case of the CS protein, the allelic polymorphism is nowhere near as old as MHC polymorphism. In the case of MSA-1, although in certain regions the polymorphism is as old as or possible older than the oldest MHC polymorphism, it appears likely that T cell epitopes are drawn from more recently homogenized regions. Indeed, it may prove to be generally true that T cell epitopes may generally be taken from fairly conserved regions, in which case polymorphisms that have evolved in such regions with the purpose of evading host recognition may be relatively short-lived in an evolutionary sense. It is in the epitopes recognized by immunoglobulins that we may then find the most ancient and long-persistent of parasite polymorphisms.

SUMMARY

The malaria parasite *P. falciparum* is a virulent and widespread parasite of humans. Contrary to previous claims, available molecular evidence does not support the idea that this parasite has only recently been transferred to humans from another vertebrate host. Analysis of DNA sequences from the CS protein and MSA-1 loci of *P. falciparum* reveals that positive selection has acted to diversify regions that may interact with the host immune system; thus, polymorphism at these loci appears to be maintained by balancing selection favoring evasion of host immune recognition. CS protein polymorphism in *P. falciparum* is probably not very old (perhaps 2 million years). MSA-1 alleles have frequently been involved in events of interallelic genetic exchange of two types: (1) nonreciprocal events which have homogenized certain regions across allelic lineages; (2) reciprocal events which have exchanged DNA segments between alleles belonging to different lineages. In the most highly divergent region, these alleles fall into two groups; divergence time between the two is estimated by DNA sequence comparison at about 35 million years ago. However, the available evidence suggests that peptides derived from *Plasmodium* antigens that are presented by the vertebrate MHC derive from more conserved regions. If polymorphism has evolved in such regions as an adaptation to evade binding by the MHC, it has evolved relatively recently, in contrast to what is observed in the case of the most highly divergent regions of the *P. falciparum* MSA-1.

Acknowledgments

This research was supported by a grant from the National Institutes of Health.

REFERENCES

Ballou, W.R., J. Rothbard, R.A. Wirtz, R.W. Gore, I. Schneider, M.R. Hollingdale, R.L. Beaudoin, W.L. Maloy, L.H. Miller, and W.T. Hockmeyer, 1985. *Science* **228**: 996-999.

Cameron, T.W.M., 1956. *Parasites and Parasitism*. Methuen, London.

Coatney, G.R., W.E. Collins, M. Warren, and P.G. Contacos, 1971. *The Primate Malarias*. U.S. Dept. of Health, Education, and Welfare, Bethesda.

Cochrane, A.H., F. Santoro, V. Nussenzweig, R.W. Gwadz, and R.S. Nussenzweig, 1982. *Proc. Natl. Acad. Sci. U.S.A.* **79**: 5651-5655.

Crisanti, A., H.-M. Müller, C. Hilbich, F. Sinigaglia, H. Matile, M. McKay, J. Scaife, K. Beyrenther, and H. Bujard, 1988. *Science* **240**: 1324-1326.

Cross, G.A.M., 1990. *Annu. Rev. Immunol.* **8**: 83-110.

De Groot, A.S., A.M. Johnson, W.L. Maloy, I.A. Quakyi, E.M. Riley, A. Menon, S.M. Banks, J.A. Berzotsky, and M.F. Good, 1989. *J. Immunol.* **142**: 4000-4005.

de la Cruz, V.F., A.A. Lal, and T.F. McCutchan, 1987. *J. Biol. Chem.* **262**: 11935-11939.

Doherty, P.C. and R.M. Zinkernagel, 1975. *Nature* **256**: 50-52.

Enea, V. and D. Arnot, 1988. In *Molecular Genetics of Parasitic Protozoa*, edited by M.J. Turner and D. Arnot, pp. 5-11. Cold Spring Harbor Laboratory, Cold Spring Harbor, New York.

Galinski, M.R., D.E. Arnot, A.H. Cochrane, J.W. Barnwell, R.S. Nussensweig, and V. Enea, 1987. *Cell* **48**: 311-319.

Garnham, P.C.C., 1988. In *Malaria: Principles and Practice of Malariology*, Vol. 1, edited by W.H. Wernsdorfer and I. McGregor, pp. 61-96, Churchill Livingstone, Edinburgh.

Gyllensten, U.B. and H.A. Erlich, 1989. *Proc. Natl. Acad. Sci. U.S.A.* **86**: 9986-9990.

Hill, A.V.S., C.E.M. Allsopp, D. Kwiatkowski, N.M. Anstey, P. Twumasi, P.A. Rowe, S. Bennett, D. Brewster, A.J. McMichael, and B.M. Greenwood, 1991. *Nature* **352**: 595-600.

Hoffman, S.L., C.N. Oster, C. Mason, J.C. Beier, J.-A. Sherwood, W.R. Ballou, M. Mugambi, and J.D. Chulay, 1989. *J. Immunol.* **142**: 1299-1303.

Howard, J.G., 1987. In *Evolution and Vertebrate Immunity*, edited by G. Kelsoe and D.H. Schulze, pp. 397-427, University of Texas Press, Austin.

Hughes, A.L., 1991. *Genetics* **127**: 345-353.

Hughes, A.L., 1992. *Mol. Biol. Evol.* **9**: 381-393.

Hughes, A.L. and M. Nei, 1988. *Nature* **335**: 167-170.

Hughes, A.L. and M. Nei, 1989. *Proc. Natl. Acad. Sci. U.S.A.* **86**: 958-962.

Hughes, A.L., T. Ota, and M. Nei, 1990. *Mol. Biol. Evol.* **7**: 515-524.

Kemp, D.J., R.L. Coppel, and R.F. Andrews, 1987. *Annu. Rev. Microbiol.* **41**: 181-208.

Klein, J., 1986. *Natural History of the Major Histocompatibility Complex*. John Wiley & Sons, New York.

Klein, J., 1991. *Human Immunol.* **30**: 247-258.

Kreier, J.P. and J.D. Baker, 1987. *Parasitic Protozoa*. Allen and Unwin, Boston.

Kumar, S., L.H. Miller, I.A. Quakyi, D.B. Keisler, R.A. Houghten, W.L. Maloy, B. Moss, J.A. Berzotski, and M.F. Good, 1988. *Nature* **334**: 258-260.

Lewis, A.L., 1989. *Mol. Biochem. Parasitol.* **36**: 271-282.

Li, W.-H., 1989. *Mol. Biol. Evol.* **6**: 424-435.

Li, W.-H., M. Gouy, P.M. Sharp, C. O'hUigin, and Y.W. Yang, 1990. *Proc. Natl. Acad. Sci. U.S.A.* **87**: 6703-6707.

Livingstone, F.B., 1958. *Am. Anthropol.* **60**: 531-561.

May, R.M., 1985. In *Ecology and Genetics of Host-Parasite Interactions*, edited by D. Rollinson and R.M. Anderson, pp. 243-262. Academic Press, London.

Mayer, W.E., M. Jonker, D. Klein, P. Ivanyi, G. van Seventer, and J. Klein, 1988. *EMBO J.* **7**: 2765-2774.

McCutchan, T.F., J.B. Dame, L.H. Miller, and J. Barnwell, 1984. *Science* **225**: 808-811.

Miyata, T., S. Miyazawa, and T. Yasunaga, 1979. *J. Mol. Evol.* **12**: 219-236.

Molineaux, L., 1988. In *Malaria: Principles and Practice of Malariology*, Vol. 2, edited by W.H. Wernsdorfer and I. McGregor, pp. 913-998. Churchill Livingstone, Edinburgh.

Monos, D.S., W.A. Tekolf, S. Shaw, and H.L. Cooper, 1984. *J. Immunol.* **132**: 1379-1385.

Nei, M. and T. Gojobori, 1986. *Mol. Biol. Evol.* **3**: 418-426.

Nei, M. and L. Jin, 1989. *Mol. Biol. Evol.* **6**: 290-300.

Nelson, K., T.S. Whittam, and R.K. Selander, 1991. *Proc. Natl. Acad. Sci. U.S.A.* **88**: 6667-6671.

Saitou, N. and M. Nei, 1987. *Mol. Biol. Evol.* **4**: 406-425.

Sette, A., S. Buus, S. Colon, C. Miles, and H.M. Grey, 1988. *J. Immunol.* **141**: 45-48.

Smith, N.H., P. Beltran, and R.K. Selander, 1990. *J. Bacteriol.* **172**: 2209-2216.

Takahata, N. and M. Nei, 1990. *Genetics* **124**: 967-978.

Tibayrenc, M., F. Kjellberg, and F.J. Ayala, 1990. *Proc. Natl. Acad. Sci. U.S.A.* **87**: 2414-2418.

Tibayrenc, M., F. Kjellberg, J. Arnaud, B. Oury, S.F. Brenière, M.-L. Dardé, and F.J. Ayala,

1991. *Proc. Natl. Acad. Sci. U.S.A.* **88**: 5129-5133.

Vergara, U., R. Gwadz, D. Schlesinger, V. Nussenzweig, and A. Feriera, 1986. *Mol. Biochem. Parasitol.* **14**: 283-292.

Waters, A.D., D.G. Higgins, and T.F. McCutchan, 1991. *Proc. Natl. Acad. Sci. U.S.A.* **88**: 3140-3144.

Weiss, W.R., M. Sedegah, R.L. Beaudoin, L.H. Miller, and M.F. Good, 1988. *Proc. Natl. Acad. Sci. U.S.A.* **85**: 573-576.

Yoshida, N., R.S. Nussenzweig, P. Potochjak, V. Nussenzweig, and M. Aikawa, 1980. *Science* **207**: 71-73.

Balancing Selection at *HLA* Loci

YOKO SATTA

Department of Population Genetics,
National Institute of Genetics, Mishima 411, Japan

One remarkable characteristic of polymorphism at the major histocompatibility complex (*Mhc*) loci is the unusually high extent of sequence differences. The average nucleotide diversity (π) at the *HLA* (human *Mhc*) loci is 2% to 4%, while that for other nuclear loci ranges from 0.02% to 0.2 % (Li and Sadler, 1991; Nei and Hughes, 1991 and references therein). The π value even for the human mitochondrial DNA is still only 0.4%, of which the mutation rate (per site per year) is estimated to be 5 to 10 times higher than that of the nuclear DNAs (Cann *et al.*, 1987; Wilson *et al.*, 1985).

Neutrality (Kimura, 1968, 1983) is inconsistent with the *trans*-species mode of polymorphism at the *Mhc* loci: Certain alleles of one species are generally more similar to certain alleles of another species than they are to conspecific alleles (Klein, 1980; Figueroa *et al.*, 1988; Lawlor *et al.*, 1988; Mayer *et al.*, 1988). Such tremendous polymorphism is due to the fact that most of the *Mhc* alleles diverged long before species divergence. Without selection, there are two explanations for this pattern: a high mutation rate and population structure. However, a high mutation rate makes allelic turnover rates fast so that allelic lineages cannot persist for tens of million years (Crow, 1972; Kimura and Ohta, 1973; Takahata, 1990a; Takahata and Nei, 1990). A strongly structured population can maintain very old alleles (Takahata, 1988, 1991a), but such structure affects the persistence time of allelic lineages at other loci as well. This expectation is inconsistent with most of the available data (Cann *et al.*, 1987; Xiong *et al.*, 1991; Li and Sadler, 1991). In addition to the high non-synonymous substitution rate at the peptide-binding region (PBR) (Hughes and Nei, 1988; see below), these considerations have suggested that some type of selection must be invoked in the *Mhc* polymorphism (Takahata and Nei, 1990; Takahata, 1990a).

Information on the three-dimensional structure of an *Mhc* molecule has provided a way to look for evidence of adaptive selection operating on this gene. In 1987, Bjorkman and her coworkers determined the structure of an *HLA* Class I molecule and identified the amino acid sites of the pocket into which antigens bind (Bjorkman *et al.*, 1987). Hughes and Nei (1988, 1989) examined the distance (the number of nucleotide substitutions per site) at the synonymous and nonsynonymous sites in the PBR as well as in other regions. They demonstrated that the distance at the nonsynonymous sites in the PBR is about 3 times as large as that at the synonymous sites and that such a large excess of nonsynonymous changes is restricted to the PBR. They therefore concluded that Darwinian selection (overdominance type) has operated on nonsynonymous changes in the PBR (see also Maruyama and Nei, 1981). However, the detailed mode of selection, as well as the biological mechanism, is still controversial (Doherty and Zinkernagel, 1975; Klein, 1986; Hughes and Nei, 1988; Takahata and Nei, 1990; Takahata, 1990a; Nei and Hughes, 1991; Hedrick *et al.*, 1991).

When we analyze data of *Mhc* polymorphism, population genetic consider-ations are essential. A model of symmetric balancing selection has proven useful for the analysis. Takahata *et al.* (1992) have recently examined various predictions of this model and showed the consistency with *HLA* data. However, their analysis was largely on the synonymous and PBR nonsynonymous changes. Here, I focus on the non-PBR nonsynonymous changes and further examine the consistency of data with the model. Also, since the calibration of synonymous substitution rates is important for understanding the *Mhc* evolu-tion, three Class II loci (*DRB1*, *DQB1*, and *DQA1*) are used for this purpose.

HLA DATA ANALYSIS

1. *Synonymous Substitution Rate*

The *trans*-specific mode of *Mhc* polymorphism makes the calibration of substitution rates difficult. When we compare nucleotide sequences of *Mhc* alleles among primates, the distance between human and chimpanzee alleles is as large as that between human (or chimpanzee) and macaque alleles. In the case of the *DRB1* locus, the former ranges from 5.0% to 9.5%, while the latter ranges from 4.7% to 11.1%. This could happen if allelic lineages in humans and chimpanzees diverged long before these two species diverged. Therefore, the ordinary method for calibrating the substitution rate (see Nei, 1987 for review), which assumes species divergence to be identical to gene divergence, tends to overestimate the substitution rate of *Mhc* genes. If we instead compare distantly related species, the species divergence time can be used as the gene divergence time (Pamilo and Nei, 1988; Takahata, 1989). However, multiple hit substitutions may be extensive so that it is often difficult to infer accurately the actual number of substitutions in such a comparison. If this happens, the rate is, of course, underestimated (Nei, 1987).

There is a way, however, to overcome both difficulties mentioned above when a number of alleles sequenced are available from various species (Satta *et al.*, 1991). The idea is simple: if a number of interspecific comparisons of alleles are possible in a sample, there might be pairs of interspecific alleles whose divergences occurred close to species divergences. Such pairs are expected to show the smallest substitution rate among interspecific comparisons. Since the time difference between allelic and species divergences is necessarily a random variable (Tajima, 1983; Takahata and Nei, 1985; Neigel and Avise, 1986; Nei, 1987; Takahata, 1989; Satta *et al.*, 1991), it is necessary to examine many combinations of interspecific alleles. It is also necessary to show that the

TABLE I

GC% at the Third Codon Positions, and the Mean and Standard Error ($\pm$) of Synonymous Substitution Rate

Gene	GC% at the 3rd	Substitution rate (10^{-9}/site/year)	*n* site
TPI	64.4	2.5±0.6	176.0
α-Antitrypsin	67.7	3.1±0.5	275.2
Pepsinogen A	78.9	2.1±0.4	274.7
CGβ	81.6	2.3±0.6	129.8
ApoAI	85.6	1.6±0.4	181.3
ApoE	88.1	2.0±0.4	234.8
DRB1	72.4	1.2±0.4	165.8
DQB1	86.3	1.6±1.6	40.0
DQA1	57.0	1.2±0.9	36.8

n site is the number of synonymous sites compared. *TPI*: triose phosphate isomerase. *CGβ*: chorionic gonadotropin-*β*. *Apo*: apolipoprotein.

Source of data: all of the nucleotide sequences of non-*Mhc* genes are taken from Genebank. *DRB1*: Satta *et al.* (1991) and references therein. *DQB1*: Gyllensten *et al.* (1990) and references therein. *DQA1*: Gyllensten and Erlich (1989) and references therein.

smallest rate computed from such pairs of alleles is not an artifact of the rate-slowdown in the lineages leading to the species involved (Satta *et al.*, 1991).

The estimated synonymous rates at three Class II loci (*DRB1*, *DQB1*, and *DQA1*) are 1.2–1.6×10^{-9} per site per year (see legend of Table I for the references of the sequences used). These rates were estimated between hominids and Old World monkeys for *DRB1* and *DQA1*, and between gorilla and human for *DQB1*. With a New World monkey or cattle sequence as a reference, the relative rate test (Wilson *et al.*, 1977; Wu and Li, 1985) shows no significant ($p > 0.1\%$) slowdown in either hominids or catarrhine lineages (see also Easteal, 1991). Thus the estimated rates are not an artifact of the rate-slowdown in the lineages concerned. Unfortunately, this method could not be applied to Class I genes for which the sequence data in primates have been limited to humans and chimpanzees.

The estimated rate of three Class II genes is slower than that of any known gene (Li *et al.*, 1985; Bulmer *et al.*, 1991). The slow synonymous substitution rate of *Mhc* genes was previously suggested and the cause was attributed to the high GC content at the third codon positions, *e.g.*, 72% GC for mammalian *DRB* genes (Hayashida and Miyata, 1983; Hughes and Nei, 1988, 1989). To

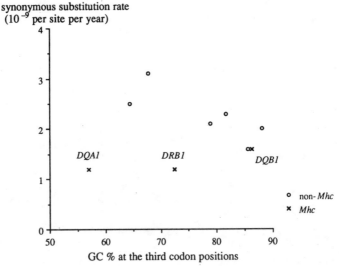

Fig. 1. Relationships between GC content at the third codon positions and synonymous rate. Estimated synonymous rates are taken from Table I. Although the GC content differs significantly among *Mhc* genes, the rates are similar to each other and lower than those of non-*Mhc* genes.

examine the relationship between GC contents and synonymous substitution rates (Wolfe *et al.*, 1989; Satta and Takahata, 1990; Bulmer *et al.*, 1991), six non-*Mhc* genes with high GC contents at the third codon positions (64% to 88%) were chosen and the synonymous substitution rate of each gene was estimated from the comparison between humans and Old World monkeys (Table I). The estimated rate ranges from 1.6×10^{-9} to 3.1×10^{-9} per site per year and, in fact, shows a negative correlation with the GC content at the third codon positions (Fig. 1). However, the rate at the Class II loci is rather uniform, independent of the GC content. An alternative explanation for the slow synonymous substitution rate is therefore the lowered mutation rate in the *Mhc* coding region. If this is the case, the substitution rate in nearby noncoding regions, such as introns, must also be lowered.

2. *Nonsynonymous Changes and Domain Transfer*

The high positive correlation between the number of synonymous changes (S) and that of non-PBR nonsynonymous changes (N) in all pairwise comparisons of sampled alleles indicates that the linkage between these sites is generally tight (Takahata *et al.*, 1992). There are $i(i-1)/2$ pairs of N and S for a sample of i alleles. The correlation coefficient (r) at each of the *HLA-A*, *-B*, *-C*, and *-DRB1* loci becomes 0.78, 0.55, 0.88, and 0.68, respectively. The correlation between S and B (the number of PBR nonsynonymous changes per gene) at each locus is also as high as that between S and N. Since the synonymous, PBR and non-PBR nonsynonymous sites are linked and dispersed throughout the entire gene, large values of r can be anticipated. Nevertheless, if recombination, gene conversion, acceleration of PBR nonsynonymous rates in particular allelic lineages, or all three of these occur, r becomes small and the distribution of N in all pairwise comparisons becomes broad.

A test was carried out to examine whether the broad distribution of non-PBR nonsynonymous changes N for a pair of alleles can be expected solely by chance. The expected proportion (p) of N relative to synonymous and non-PBR nonsynonymous changes was calculated by $p = N_P/(S_P + N_P)$, where subscript P stands for the mean number of substitutions per gene taken over all pairwise comparisons. The distribution of N for a given pair of alleles was assumed to be binomial with parameters p and the $N+S$ value, from which the 95% confidence limit was set. The distribution of N is so broad as to be by and large compatible with the data.

However, some allelic pairs in fact exhibit unusual values of N and S, and in two cases, small values of r were attributed to particular alleles (*i.e.*, allele *2401* at the *A* locus and *1001* at the *DRB1* locus). The correlation at the *B* locus

A1

	α1	α2
	* * *****	**** **********
0101	TACCGCAGGACCAA	CCCGGGAATAGCCGTCGCGA
1101	ACCCGCGGGGTGA	CCCGGGAAGAGCCACCGGTG
3001	TAATAAGATGGTGC	TTGAACGCGGTGTTCACGTA

A2

	α1	α2	α3+TM+CYT
	* * *******	*** *** ******	
0201	TTAATGGGGACCC	GGGTTACTACATTTACA	GAGTGTGAACCTCTTCT
6901	CACCGCGACTGGC	GGGTTACTACATTTACA	GAGTGTGAACCTCTTCT
6801	CACCGCGACTGGC	ATGTGGGGACATTGACA	GAGTGTGAATCTCTTCT
1101	CACCGCAGGTGGA	ATACGGGGCTGCCACGG	CGACCCAGGCGCTCCTA

A3

	α1	α2	α3+TM+CYT
	** *******	* * ******	
0301	TTCGGGAGGCTG	AGGCAAAGCTT	CAACCCAGCGGGTGGCAACTTACTGC
2901	ACTAGGTCGCCA	GCGTCGTGCGT	GAGTGTGATTCGCAATTGTGCTTCAT
3101	ACCGATAGCTTG	GGATCGTGCGT	CAGTGTGATTCCCGATTGTGCTCTAT
3301	ACCGGGGACTTG	GGATCGTCGGC	CGGTGTGACTCCCGATTGTGCTTCAT

B1

	α1	α2	α3+TM+CYT
	*****************	** **	
3501	CACACATTCAAAAAGACTGG	AGCCCGCC	CGTGGAAC
3502	CACACATTCAAAAAGACTGG	AGCCTAAC	CGTGGAAC
5701	TCGGGGAAGGTCGGATGCTC	GTGGCGCG	TACAAGGG
5801	CAGGGGAAGGTCGGATGCTC	AGCCCGCC	CGTGGAAC

B2

	α1	α2	α3+TM+CYT
	* ** ** *****	****** ** ******	
4101	CGAACGTAGAAGGCAACATAGGACTGG	CTTGGGATCTATCGCGGACACGG	CCTA
4201	TTGTACAGAGGACAGCGGGAGGACTGG	CCCTCCACCTATCGCGGACACGG	CCTA
4402	TGAACGTAGAAGGCAACATAACCGCTC	TCATCGACTTGGAGCCGACCTCC	TTCA
4701	TGAACGTAGAAGGCAACATGAGCCTTC	CCCTCGTTTCCGACGGCTGGAGC	CCTG

B3

	α1	α2	α3+TM+CYT
	* ** *****************	**** *** *******	
1801	CTGTAGCAGCACGAACTCCACACATAGACTGG	CCTCCGGCGCCAGCAGTCTACCGCCG	TCCAAACC
4601	TGAGAACAGTCCAGGGAGACGCGGGTGACTGG	CCTCCGGCGCCAGGAGATGCTTGCCG	TCCAAACC
4001	CGAACAGTAAAGAGGGTCCACACATAGACTGG	CCTCCGGCAACCCGTTTCTGATCAGT	CTTGGGTG
4901	CGAACAGTAAAGAGGGTCCACACATAATGCTC	TTGGTCCTATAAGCAGACTCTTGCCG	TCCAAACC

DRB

	β1	β2
	** ** * **** ** ** * ***** *******	
1301	AGTGATCTCCCGTGTCGCTATCGAATGATAAAGACGACGCTACTGC	ATACTTACGCCA
0701	CACTGCGGGAAACATCGACTTTGTTAATCCAGGACAGGCAGTGGTC	GGCTCCGGATTG
0901	CATAACGGAATGCGATCCGGATAAAAGTTCTGCGGAGCGAGTGGTA	GGCTCCGGATTG

is also rather small, but there appears to be not one but several alleles responsible. The low correlation of *A* 2401* results from an excess of *N* in the α1 (exon 2) and α3 (exon 4) domains. Such an excess is due to the appearance of motif TCGCGCTCC (shared only by *A* 2501* and *3201*) in the α1 domain and many unique nonsynonymous changes in the α3 domain. This motif in the α1 domain has also been reported by Hedrick *et al.* (1991). *DRB1* 1001* has six unique codons which occur both inside and outside the PBR in the β1 domain, but only one in the β2 domain. The correlation between *S* and *N* within each of two domains is higher than that in the entire gene.

These results forced more careful examination of the data from domain to domain. A test similar to the above was again applied to look at the correlation among domains. Here, the total number (*T*) of nucleotide changes in each domain was used, and the *p* was defined this time as the proportion of the mean pairwise number of nucleotide changes in each domain to that in all the domains. Anomalous correlations were found in comparisons of (A1) *A* 0101*, *1101*, and *3001* relative to other alleles (*e.g.*, *A* 0201*), (A2) *A* 0201* (including closely related *0203*, *0205*, *0206*, and *0210*), *1101*, and *6901* (*6801* and *6802* as well), (A3) *A* 0301*, *2901*, *3101*, and *3301*, (B1) *B* 3501*, *3502*, *5701*, and *5801*, (B2) *B* 4101*, *4201*, *4402*, and *4701*, (B3) *B* 1801*, *4001*, *4601*, and *4901*, and (DRB) *DRB1* 0701* and *0901* relative to *DRB1* 1301*.

Figure 2 shows the nucleotide changes among alleles in each of the comparisons (A1), (A2), (A3), (B1), (B2), (B3), and (DRB). Except in comparisons (B2) and (B3), boundaries of their spatial heterogeneities are consistent with boundaries of domains. In comparison (A1), the correlation between the α1 and α2 domains is high, but low between α1 + α2 and α3. This is due to the fact that there is no nucleotide change in the α3 domain, despite a number of changes in the α1 + α2 domain. The same is true for comparison (DRB) in which the β2 domain is identical between *DRB1* 0701* and *0901* but there are a large number of changes in the β1 domain. In relation to comparison (A2), Holmes and Parham (1985) noted that the nucleotide sequence of *Aw69* (*A* 6901*) is identical to that of *Aw68* (*A* 6801*) in the α1 domain, but to that of *A2(A* 0201)* in the α2 + α3 domains, concluding that *Aw69* was a hybrid molecule between *Aw68* and *A2*. However, *A* 6901* and *6801* do not show any anomalous relationship between α1 and α2 + α3. This is because these alleles are closely related to each other not only in the α1 domain but also in the other

---

←Fig. 2. Variable sites among anomalous alleles. Here and in Fig. 3, A1–3, B1–3, and DRB are the comparisons which show low correlations. In the text, these are parenthesized. Asterisks stand for sites in the PBR. TM and CYT stand for the transmembrane and cytoplasmic domain, respectively.

domains: their sequence similarity and the present statistical test simply do not demonstrate any segmental exchange of DNA sequences that might occur. It is more likely that the $\alpha 2 + \alpha 3$ domains of A^*0201 recently converted those of $A^*6801, 6901$, or the common ancestral lineage (Figs. 2 and 3). Comparison (B1) shows that three alleles are nearly identical in the $\alpha 2$ and $\alpha 3$ domains, but one of the alleles is significantly different from the other two in the $\alpha 1$ domain. Comparisons (B2) and (B3) show that the entire gene can be divided into two and four segments, respectively, with respect to sequence similarity.

As mentioned, there are three explanations for the anomalous ancestral relationships among domains; recombination, gene conversion, and acceleration of PBR nonsynonymous rates in particular allelic lineages.

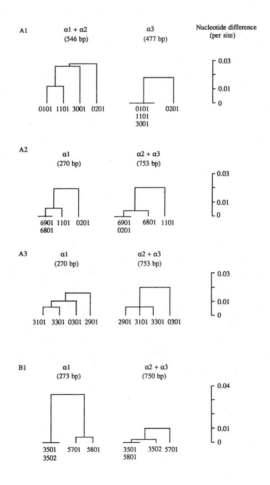

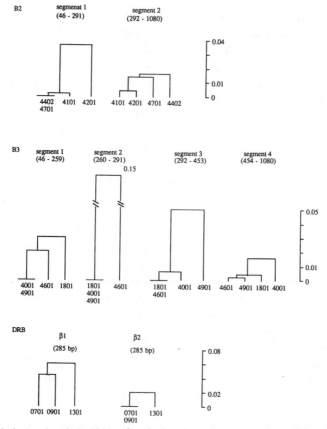

Fig. 3.   Phylogenetic relationships of each domain among anomalous alleles. A1, A2, and A3 are comparisons among such alleles at the *A* locus. B1, B2, and B3 are those at the *B* locus. DRB is for the *DRB1* locus. The number at each tip stands for allele designation taken after the WHO nomenclature. The branch length is depicted proportional to nucleotide differences per site.

In comparison (A1), the allelic lineage leading to *A*3001* has 21 single changes (10 PBR nonsynonymous, 5 synonymous, and 6 non-PBR nonsynony-mous changes, *i.e.*, $10B+5S+6N$) in the $\alpha1+\alpha2$ domains, whereas there is no single substitution in the $\alpha3$ domain. On the other hand, *A*1101* and *0101* are closely related in both regions; there are only 3 ($1B+2S$) and 9 ($8B+1N$) single changes in $\alpha1+\alpha2$, respectively, and none in $\alpha3$. The number of PBR nonsynonymous substitutions in *A*3001* is as many as that in *A*0101*. Thus the acceleration of the rate at the PBR of A^*3001 seems an unlikely explanation. If the number of neutral substitutions in $\alpha1 + \alpha2$ of A^*3001 really reflects the phylogenetic relationship, a conversion hypothesis is plausible; the $\alpha3$ domain of A^*3001 has been recently converted by that of A^*1101 or A^*0101 (Fig. 3). Similarly, comparison (A2) is an example of recent conversion, while (A3) provides a case of intragenic recombination between A^*0301 and 2901 with a break-point between $\alpha1$ and $\alpha2 + \alpha3$.

In comparison (B1), the $\alpha1$ domain sequence of B^*5801 is almost identical to that of B^*5701. Nevertheless, the $\alpha2$ and $\alpha3$ domains of B^*5801 are closely related to those of 3501 as well as 3502. Either a recent gene conversion of B^*5801 by 5701 in the $\alpha1$ domain or by 3501 in the $\alpha2 + \alpha3$ domains appears to be a reasonable explanation. In comparison (B2), phylogenetic relationship in the first half of the $\alpha1$ domain (segment 1 in Fig. 3) was different from that of the other region (segment 2): B^*4101, 4402, and 4701 are identical in the first segment which includes 11 nucleotide sites of the PBR, whereas 4101 and 4201 are very similar in the other region. Conversion of B^*4101 by either 4402 or 4701 in the former region is a likely explanation for the identical sequence including 11 nucleotide sites of the PBR in the region. In comparison (B3), however, it was difficult to infer which segment really reflects the actual ancestry of these alleles and therefore to identify what kind of molecular mechanisms is most important (Figs. 2 and 3).

In the $\beta1$ domain, in comparison (DRB), $DRB1^*0701$, 0901, and 1301 have accumulated 11 single substitutions ($5B + 2S + 4N$), 10 ($4B + 3S + 3N$), and 20 ($8B + 5S + 7N$), respectively. In the $\beta2$ domain, however, DRB^*0701 and 0901 are identical despite 12 single changes ($7S + 5N$) in the allelic lineage leading to $DRB1^*1301$. Therefore, the reconstructed trees for the $\beta1$ and $\beta2$ domains become topologically identical to each other, but they are substantially different in their branch lengths leading to $DRB1^*0701$ and $DRB1^*0901$ (Fig. 3). Although one may take this difference as evidence for gene conversion, the possibility of acceleration cannot be ruled out. Since the PBR of a Class II gene is putative (Brown et al., 1988), some non-PBR nonsynonymous sites can be targets for selection and have accumulated more changes than neutral ones. In fact, synonymous changes alone do not show any significant difference in the ancestry of the $\beta1$ and $\beta2$ domains. This possibility is further supported by the fact that the DR region is frozen in terms of recombination (Gregersen et al., 1988; O'hUigin, 1992).

Except for these comparisons, the correlation among domains is generally significantly high. In the subsequent analysis, I did not exclude any of the phylogenetically anomalous alleles and used all sequence information available at the *A*, *B*, *C*, and *DRB1* loci. Effects of including anomalous alleles will be studied elsewhere.

3. Functional Constraint on Non-PBR Nonsynonymous Sites

To estimate the degree of overall functional constraint, f, on the non-PBR nonsynonymous sites relative to the synonymous sites, we use the mean number of substitutions per gene, N_P and S_P, and define f as

$$f = \frac{N_P}{L_N} \frac{L_S}{S_P}.$$

In the above, L_N and L_S are the numbers of the non-PBR nonsynonymous and synonymous sites in the entire gene, respectively. For a neutral gene such as pseudogene, we expect $f = 1$ (Kimura, 1983), and for a completely conserved gene (no acceptable nonsynonymous substitution), $f = 0$.

Generally, N_P and S_P are small. Thus, there seems no serious statistical problem in using the mean pairwise distances to estimate f, when the constraint against the nonsynonymous changes at each locus has been kept constant during the course of evolution.

The f values of three Class I loci are 0.412 at the *A* locus, 0.385 at the *B* locus, and 0.367 at the *C* locus, while the value at the *DRB1* is 0.328 (Table II). Irrespective of loci, the degree of functional constraint is nearly 1/3.

TABLE II
Estimated Number of Common Alleles (n), Functional Constraint (f), and Ratio (γ) of the Selected Substitution Rate to the Neutral Rate

	Locus			
	A	*B*	*C*	*DRB1*
i	19	26	6	19
n^*	19	37	9	47
n	26.3 ($m=25$)	38.1 ($m=18$)	15.1 ($m=21$)	25.1 ($m=25$)
f	0.412	0.385	0.367	0.328
γ	4.7 (2.7)	7.2 (3.5)	2.9 (1.6)	8.1 (4.8)
S_P	10.8	10.2	10.2	10.3
N_P	12.0	10.3	9.9	10.5
B_P	15.4	18.3	8.4	15.1

i: the number of alleles used in this study. n^*: the number of known alleles taken from Klein (1986) for Class I (unidentified alleles are counted as one), Marsh and Bodmer (1991) for *DRB1*. The value of m which allowed estimation of n at each locus is in parenthesis in column n. The estimated ratio of γ based on B_P is given in parenthesis in column γ. S_P, N_P, and B_P are the mean pairwise number of synonymous, non-PBR and PBR nonsynonymous substitutions, respectively.

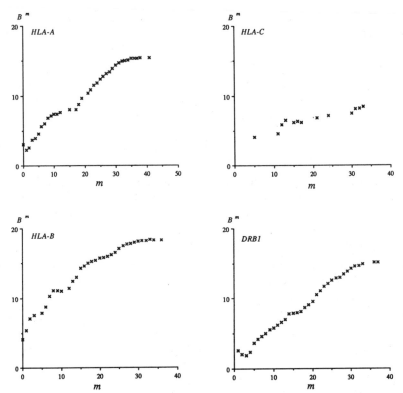

Fig. 4.   The conditional mean of PBR nonsynonymous changes (B^m) for a given m (abscissa) which is the number of synonymous and non-PBR nonsynonymous changes per allelic pair.

Although non-PBR nonsynonymous changes must be subjected to purifying selection (Kimura, 1983), the selection intensity is not particularly strong. The degree of functional constraint of non-*Mhc* genes varies greatly. It is 0.004 for the histon H4, 0.201 for the insulin C-peptide, and 0.467 for the interferon $\beta 1$, the average being 0.189 (Li *et al.*, 1985). The average f value of *HLA* molecules is about twice as large as that of other genes.

4.   The Number of Common Alleles in a Population

Under balancing selection, the allelic genealogy predicts that the mean pairwise number of substitutions at the selected sites per gene (B_P) is equal to the mean number of common alleles segregating in a population (Takahata, 1990b; Takahata *et al.*, 1992). As shown in Table II, B_P is much smaller than the experimentally observed number of alleles (n^* in the table). This appears to be mainly because multiple hit substitutions at the selected sites are not accurately inferred for pairs of distantly related alleles.

To avoid this problem, Takahata *et al.* (1992) restricted their analysis to allelic pairs whose synonymous changes are relatively small. However, the number of synonymous sites per gene is fairly small and synonymous changes are subjected to large sampling errors. It is, therefore, more desirable to use both N and S collectively. Denote by B^m the mean number of PBR non-synonymous changes among allelic pairs whose synonymous and non-PBR nonsynonymous changes are less than $m = N + S$. Figure 4 shows the conditional mean of B^m plotted against the number of m. The B^m linearly and smoothly increases as m increases up to 20 or more: the larger the number of

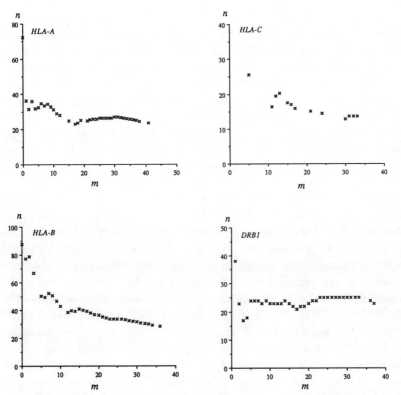

Fig. 5.   The estimated number of alleles (n) at each locus in a population. The m is the same as in Fig. 4.

the linked neutral sites, the smaller the sampling errors in the conditional mean [compare Takahata *et al.* (1992) in which only synonymous changes were used for *m*]. When *m* is larger than 30, however, the B^m levels off or is saturated. To estimate the mean number (*n*) of common alleles in a population, it is necessary to choose appropriate values of *m*. Figure 5 shows the *m* dependency of the number (*n*) of alleles estimated by the same method as in Takahata *et al.* (1992) (although this method tends to underestimate *n*, the extent of such a bias is not great; Takahata, personal communication). At the *A* and *DRB1* loci, the estimated *n* is almost constant for *m* = 25 to 35, so that *m* = 25 was chosen. Likewise, *m* = 18 for the *B* locus and *m* = 21 for the *C* locus were used. Table II shows the *n* value thus estimated at each locus. The *n* values at Class I loci are in good agreement with the observed number of alleles at these loci. The larger-than-expected value of *n* is, of course, suggestive of other unknown alleles.

However, the *n* of the *DRB1* is one half the number of already known alleles (47 detected by all available methods; see Marsh and Bodmer, 1991). There are four possible explanations for this discrepancy in the *DRB1*. (i) Usual Poisson correction methods may underestimate the actual number of substitutions at the PBR nonsynonymous sites (Jin and Nei, 1990; Takahata, 1991b). Maximum parsimony analysis shows that some sites in the PBR have accumulated seven nonsynonymous changes but others none. Such a broad distribution of the number of substitutions per site is not expected from Poisson processes (Takahata *et al.*, 1992; see also Uzzell and Corbin, 1971; Holmquist *et al.*, 1983). However, the conditional mean, B^m, is small because it is based on allelic pairs with relatively small values of *m*. When B^m is small, Poisson corrections can be accurate even though the actual process does not follow a Poisson process (Takahata, 1991b). Thus, the efficacy of the correction method might have little effect. (ii) Since the PBR in the Class II molecule is putative (Brown *et al.*, 1988), some sites outside the PBR may be targets for selection. As an extreme, we may assume all nonsynonymous substitutions to be driven by selection. The estimated number of alleles then becomes about 40. Although large, the figure is still a little smaller than the observed number of 47 (Marsh and Bodmer, 1991). Therefore, the putative PBR hypothesis for *DRB1* is plausible but may be insufficient. (iii) The allelic genealogy predicts the number of balanced common alleles. If the 47 alleles include neutral or rare alleles, the observed number must be larger than the estimated number of common alleles based on the allelic genealogy. Information on frequencies of these 47 alleles is therefore necessary. (iv) The prediction of the allelic genealogy under balancing selection is based on the model of a single locus. Unlike

Class I, a Class II *Mhc* molecule is composed of two different molecules, α and β chains (Klein, 1986). The PBR in the Class II is coded in two genes which are tightly linked (Klein, 1986). A more thorough theoretical study taking account of this fact may be needed for the evolutionary study of Class II genes (see Conclusion).

5. Rate of Nonsynonymous Substitutions at the PBR

The substitution rate is accelerated by balancing selection (Maruyama and Nei, 1981; Takahata, 1990b). The ratio (γ) of the PBR nonsynonymous to the synonymous substitution rate was estimated by Hughes and Nei (1988) based on the mean of all pairwise distances at the PBR nonsynonymous and synonymous sites. The estimated value was about 3. However, as mentioned earlier, their ratio is likely to be underestimated.

Table II shows the ratio γ, which is defined here as

$$\gamma = \frac{n}{L_B} \frac{fL_N + L_S}{N_P + S_P}$$

where L_B stands for the number of PBR nonsynonymous sites, n is the estimated number of alleles, and f is the degree of functional constraint.

The ratio γ is an index of selection intensity (Takahata *et al.*, 1992); the larger the γ, the stronger the selection. The estimated values of γ range from 3 to 8 (Table II). These are much larger than the previous estimates, suggesting more convincingly that selection has played an important role in the evolution of *Mhc* genes. The ratio even at the C locus is as large as that at the A locus. Since the number of alleles detected at the C locus is small and the extent of polymorphism is less than that at the A and B loci, the C locus has been considered to be less functional (Lawlor *et al.*, 1990). If this is the case, the estimated γ and n values indicate recent dysfunctioning of the C locus.

6. Genealogy

Under balancing selection, the genealogical relationships among different alleles are similar to those of neutral genes. The only difference is in the time scale (Takahata, 1990b). One way of examining such genealogical relationships is to look at the ratio of the largest (D_L) to the mean number (D_P) of substitutions, where D represents any different kind of site: N (non-PBR nonsynonymous), B (PBR nonsynonymous), S (synonymous), neutral ($K = N + S$), and the total ($T = K + B$). It was shown that the ratio $R_T = E\{D_L\}/E\{D_P\}$ becomes $2(1 - 1/i)$ where i is the number of sampled alleles (Takahata, 1990b; Takahata *et al.*, 1992), and $E\{\ \}$ stands for the expectation. This holds true for any of N, S, K, B, and T as long as the genealogy is not obscured by recombination, gene conversion, or extensive multiple hit substitutions.

To examine such a relationship, it is important to know the distribution of ratio $R_O = D_L / D_P$. For this purpose, I generated 10^4 independent genealogies (see Takahata, 1990b for the simulation method) and chose the i and D_P values so as to imitate actual data. The distribution of R_O is fairly strongly skewed for any values of i and D_P. The mean $E\{R_O\}$ is always larger than R_T (Table III). To examine D_P dependency of $E\{R_O\}$, D_P was changed from 5 to 100. $E\{R_O\}$ depends strongly on D_P. For $D_P = 100$ (though unrealistically large), $E\{R_O\}$ is close to R_T, while for $D_P = 5$, $E\{R_O\}$ becomes significantly larger than R_T (Fig. 6).

Table IV gives the estimates of R_O for the *HLA* loci, *HLA-A*, -*B*, -*C*, and -*DRB1*, together with the expected values of R_T. The R_O of the synonymous, non-PBR nonsynonymous changes at any of four loci is close to R_T. However, the R_O of PBR nonsynonymous changes, in particular at the *DRB1*, is larger than R_T. In order to minimize sampling errors, synonymous and non-PBR nonsynonymous (and even all) changes were summed up and the R_O was reexamined. The R_O values for K and T are only slightly different from R_T, supporting that the genealogy is similar to that generated by the model of symmetric balancing selection or the coalescence process and that the use of non-PBR nonsynonymous changes decreases the sampling errors substantially.

CONCLUSION

The present analysis of 70 alleles at four *HLA* loci indicates that the non-PBR nonsynonymous as well as synonymous changes are useful to characterize the allelic genealogy of the *HLA* loci. Tight linkage among the synonymous, non-PBR and PBR nonsynonymous sites within genes is generally confirmed. Although the mutation rate in the *Mhc* coding region may be lower than that in other regions, the functional constraint on the non-PBR nonsynonymous sites is not particularly strong. Therefore, the non-PBR nonsynonymous sites provide useful information on the allelic genealogy. For this reason, the synonymous and nonsynonymous changes were collectively used and assumed to be selectively neutral.

Based on the theory of allelic genealogy under symmetric balancing selection, I examined various predictions on *HLA* alleles and showed their good agreement with the data. However, through the present study, two problems remain unsolved. There are some allelic lineages which show significantly low correlations among domains. The cause appears to be domain transfer (*i.e.*,

TABLE III

Simulated Values of the Mean and Standard Deviation ($\pm$) of the Ratio of the Largest to the Mean Pairwise Number of Substitutions

	Locus			
	A	*B*	*C*	*DRB1*
$E\{S_L/S_P\}$	2.28 ± 0.63	2.39 ± 0.69	1.89 ± 0.38	2.32 ± 0.64
$E\{N_L/N_P\}$	2.27 ± 0.62	2.40 ± 0.73	1.88 ± 0.37	2.32 ± 0.66
$E\{B_L/B_P\}$	2.25 ± 0.64	2.27 ± 0.65	1.90 ± 0.38	2.25 ± 0.64
$E\{K_L/K_P\}$	2.17 ± 0.62	2.25 ± 0.66	1.80 ± 0.34	2.18 ± 0.62
$E\{T_L/T_P\}$	2.10 ± 0.60	2.15 ± 0.63	1.78 ± 0.33	2.12 ± 0.61

Subscripts L and P are the largest and the mean pairwise number of substitutions. Letters S, N, B, K, and T stand for the synonymous, non-PBR nonsynonymous, PBR nonsynonymous, neutral (synonymous + non-PBR nonsynonymous) and all sites, respectively. In simulation (10^4 repeats), the mean is set so as to imitate the actual sequence data at each locus (Table IV). The number (i) of sampled alleles and the expected ratio, $R_T=2(1-1/i)$ are as follows: for the A and $DRB1$ loci, $i=19$ and $R_T=1.89$; for the B iocus, $i=26$ and $R_T=1.92$; for the C locus, $i=6$ and $R_T=1.67$.

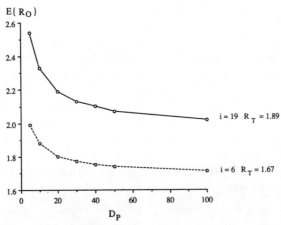

Fig. 6.   Dependency of $E\{R_0\}=E\{D_L/D_P\}$ on the mean pairwise number (D_P) of substitutions per gene. Each value of $E\{R_0\}$ (open circles in the figure) is the mean of D_L/D_P taken over 10^4 independently generated genealogies. When D_P is large, $E\{R_0\}$ is close to $R_T = E\{D_L\}/E\{D_P\}$ for any sample size i. The standard deviation of R_0 is about one third of $E\{R_0\}$ for each case.

intragenic recombination or gene conversion) or accelerated PBR nonsynonymous substitutions in particular allelic lineages. It remains to be resolved which is most important. Furthermore, whichever the main cause is,

TABLE IV
Observed Ratio of the Largest to the Mean Number of Substitutions at Each Locus

	Locus			
	A	B	C	DRB1
i	19	26	6	19
$R_T = 2(1 - 1/i)$	1.89	1.92	1.67	1.89
S_L/S_P	1.95	1.94	1.69	1.82
	(10.8)	(10.2)	(10.2)	(10.3)
N_L/N_P	1.96	1.78	1.70	1.87
	(12.0)	(10.3)	(9.9)	(10.5)
B_L/B_P	1.86	1.86	1.86	2.42
	(15.4)	(18.3)	(8.4)	(15.1)
K_L/K_P	1.80	1.74	1.67	1.75
	(22.7)	(20.5)	(20.1)	(20.8)
T_L/T_P	1.75	1.63	1.65	1.91
	(38.2)	(38.8)	(28.5)	(35.9)

Symbols are the same as those in Table III. The mean pairwise number of substitutions is given in parenthesis.

it will affect the estimates of population parameters. This problem is left also for future analysis.

It was suggested that the mode of selection on Class II genes may be different from that on Class I genes. The Class II molecule is composed of two different gene products and selection must have operated on the heterodimer as a whole. A two locus theory of balancing selection is therefore required to account for the evolution of Class II molecules. A preliminary study (Takahata, personal communication) suggests that balancing selection is less efficient for Class II than for Class I, especially in terms of the nonsynonymous substitution rate at the PBR. If this turns out to be significant and general, the number of common alleles at the *DRB1* locus might be underestimated by Eq. 12 in Takahata *et al.* (1992).

SUMMARY

The *trans*-specific polymorphism at the *Mhc* loci appears to be due to balancing selection acting at the nonsynonymous sites in the PBR. It must be manifest not only at the PBR nonsynonymous sites but at the linked synonymous and non-PBR nonsynonymous sites. Available DNA sequence data of *HLA* (human *Mhc*) loci were examined concerning the synonymous (substitution) rate, the degree of functional constraints (f) against the non-PBR nonsynonymous sites, the PBR nonsynonymous (substitution) rate, and the genealogical consistency among the synonymous, non-PBR and PBR non-synonymous sites as well as among different domains at each locus. (1) The synonymous rate of *HLA* genes was slowest among known genes. The high GC content at the third codon positions is not responsible for the slow rate, suggesting that the mutation rate itself is low in the *HLA* region. (2) The estimated value of f was about $1/3$ irrespective of loci so that the non-PBR nonsynonymous sites of the *HLA* genes are only moderately conserved. (3) The PBR nonsynonymous rate was 3 to 8 times faster than the synonymous one. This rate is faster than the previous estimate, confirming operation of fairly strong selection. (4) There were high positive correlations among nucleotide changes at the synonymous, non-PBR and PBR nonsynonymous sites. However, some alleles exhibited low correlations among different domains. Three possibilities, accelerated PBR nonsynonymous rates in particular allelic lineages, gene conversion and intragenic recombination, were discussed. For the Class I loci, the allelic genealogy agreed well with that expected from the model. In particular, the estimated number of alleles at a locus was close to that observed serologically or electrophoretically. For the *DRB1* locus, however, the number was about one half as many, and possible causes for this discrepancy were discussed.

Acknowledgments

I am grateful to Dr. Naoyuki Takahata for his encouragement and helpful comments during this work and to Dr. Jan Klein for his interest and for sending me a compiled data set.

REFERENCES

Bjorkman, P.J., M.A. Saper, B. Samraoui, W.S. Bennett, J.L. Strominger, and D.C. Wiley, 1987. *Nature* **329**: 512–518.

Brown, J.H., T. Jardetzky, M.A. Saper, B. Samraoui, P.J. Bjorkman, and D.C. Wiley, 1988. *Nature* **332**: 845–850.

Bulmer, M., K.H. Wolfe, and P.M. Sharp, 1991. *Proc. Natl. Acad. Sci. U.S.A.* **88**: 5974–5978.

Cann, R.L., M. Stoneking, and A.C. Wilson, 1987. *Nature* **325**: 31–36.

Crow, J.F., 1972. *J. Hered.* **63**: 306–316.

Doherty, P.C. and R.M. Zinkernagel, 1975. *Nature* **256**: 50–52.

Easteal, S., 1991. *Mol. Biol. Evol.* **8**: 115–127.

Figueroa, F., E. Günther, and J. Klein, 1988. *Nature* **335**: 265–267.

Gregersen, P.K., H. Kao, A. Nunez-Roldan, C.K. Hurley, R.W. Karr, and J. Silver, 1988. *J. Immunol.* **141**: 1365–1368.

Gyllensten, U.B. and H.A. Erlich, 1989. *Proc. Natl. Acad. Sci. U.S.A.* **86**: 9986–9990.

Gyllensten, U.B., D. Lashkari, and H.A. Erlich, 1990. *Proc. Natl. Acad. Sci. U.S.A.* **87**: 1835–1839.

Hayashida, H. and T. Miyata, 1983. *Proc. Natl. Acad. Sci. U.S.A.* **80**: 2671–2675.

Hedrick, P.W., W. Klitz, W.P. Robinson, M.K. Kuhner, and G. Thomson, 1991. In *Evolution at the Molecular Level*, edited by R.K. Selander, A.G. Clark, and T.S. Whittam, pp. 248–271. Sinauer, Sunderland, Massachusetts.

Holmquist, R., M. Goodman, T. Conroy, and J. Czelusniak, 1983. *J. Mol. Evol.* **19**: 437–448.

Holmes, N. and P. Parham, 1985. *EMBO J.* **4**: 2849–2854.

Hughes, A.L. and M. Nei, 1988. *Nature* **335**: 167–170.

Hughes, A.L. and M. Nei, 1989. *Proc. Natl. Acad. Sci. U.S.A.* **86**: 958–962.

Jin, L. and M. Nei, 1990. *Mol. Biol. Evol.* **7**: 82–102.

Kimura, M., 1968. *Nature* **217**: 624–626.

Kimura, M., 1983. *The Neutral Theory of Molecular Evolution.* Cambridge University Press, Cambridge.

Kimura, M. and T. Ohta, 1973. *Genetics* **75**: 199–212.

Klein, J., 1980. In *Immunology 80*, edited by M. Fougereau and J. Dausset, pp. 239–253. Academic Press, London.

Klein, J., 1986. *Natural History of the Major Histocompatibility Complex.* John Wiley & Sons, New York.

Lawlor, D.A., J. Zemmour, P.P. Ennis, A.P. Jackson, and P. Parham, 1988. *Nature* **335**: 268–271.

Lawlor, D.A., J. Zemmour, P.P. Ennis, and P. Parham, 1990. *Annu. Rev. Immunol.* **8**: 23–63.

Li, W.-H. and L.A. Sadler, 1991. *Genetics* **129**: 513–523.

Li, W.-H., C.-C. Luo, and C.-I. Wu, 1985. In *Molecular Evolutionary Genetics*, edited by R. J. Macintyre, pp. 1–94. Plenum, New York.

Marsh, S.G.E. and J. Bodmer, 1991. *Immunogenetics* **33**: 321–334.

Maruyama, T. and M. Nei, 1981. *Genetics* **98**: 441–459.

Mayer, W.E., M. Jonker, D. Klein, P. Ivanyi, G. van Seventer, and J. Klein, 1988. *EMBO J.* **7**: 2765–2774.

Nei, M., 1987. *Molecular Evolutionary Genetics.* Columbia University Press, New York.

Nei, M. and A.L. Hughes, 1991. In *Evolution at the Molecular Level*, edited by R.K. Selander, A.G. Clark, and T.S. Whittam, pp. 222–247. Sinauer, Sunderland, Massachusetts.

Neigel, J.E. and J.C. Avise, 1986. In *Evolutionary Processes and Theory*, edited by S. Karlin and E. Nevo, pp. 515–534. Academic Press, Orlando.

O'hUigin, C., 1992. In *Proceedings of the 17th Taniguchi International Symposium on Biophysics*, edited by N. Takahata, pp. 265–279.

Pamilo, P. and M. Nei, 1988. *Mol. Biol. Evol.* **5**: 568–583.

Satta, Y. and N. Takahata, 1990. *Proc. Natl. Acad. Sci. U.S.A.* **87**: 9558–9562.

Satta, Y., N. Takahata, C. Schönbach, J. Gutknecht, and J. Klein, 1991. In *Molecular Evolution of the Major Histocompatibility Complex*, edited by J. Klein and D. Klein, pp. 51–62. Springer-Verlag, Heidelberg.

Tajima, F., 1983. *Genetics* **105**: 437–460.

Takahata, N., 1988. *Genet. Res. Camb.* **52**: 213–222.

Takahata, N., 1989. *Genetics* **122**: 957–966.

Takahata, N., 1990a. In *Population Biology of Genes and Molecules*, edited by N. Takahata and

J.F. Crow, pp. 267–286. Baifukan, Tokyo.

Takahata, N., 1990b. *Proc. Natl. Acad. Sci. U.S.A.* **87**: 2419–2423.

Takahata, N., 1991a. *Genetics* **129**: 585–595.

Takahata, N., 1991b. *Proc. R. Soc. Lond. B* **243**: 13–18.

Takahata, N. and M. Nei, 1985. *Genetics* **110**: 325–344.

Takahata, N. and M. Nei, 1990. *Genetics* **124**: 967–978.

Takahata, N., Y. Satta, and J. Klein, 1992. *Genetics* **130**: 925–938.

Uzzell, T. and K.W. Corbin, 1971. *Science* **172**: 1089–1096.

Wilson, A.C., S.S. Carlson, and T.J. White, 1977. *Annu. Rev. Biochem.* **46**: 573–639.

Wilson, A.C., R.L. Cann, S.M. Carr, M. George, U.B. Gyllensten, K.M. Helm-Bychowski, R.G. Higuchi, S.R. Palumbi, E.M. Prager, R.D. Sage, and M. Stoneking, 1985. *Biol. J. Linnean Soc.* **26**: 385–400.

Wolfe, K.H., P.M. Sharp, and W.-H. Li, 1989. *Nature* **337**: 283–285.

Wu, C.-I. and W.-H. Li, 1985. *Proc. Natl. Acad. Sci. U.S.A.* **82**: 1741–1745.

Xiong, W., W.-H. Li, I. Posner, T. Yamamura, A. Yamamoto, A.M. Gotto, Jr., and L. Chan, 1991. *Am. J. Hum. Genet.* **48**: 383–389.

Evolution of the Mouse *t* Haplotype

TAKASHI MORITA,[*1] HIROSHI KUBOTA,[*1]
YOKO SATTA,[*2] AND AIZO MATSUSHIRO[*1]

*Department of Microbial Genetics, Research Institute for Microbial Diseases, Osaka University, Suita 565[*1] and National Institute of Genetics, Mishima 411,[*2] Japan*

The mouse *t* haplotype (Silver, 1985; Frischauf, 1985; Committee for Mouse Chromosome 17, 1991) is widely occurring in wild populations at frequencies of 10–40% (Klein *et al.*, 1984; Ruvinsky *et al.*, 1991). The spread is thought to depend on the transmission ratio distortion (TRD) which transmits the *t* haplotype 10 to 100 times more frequently than wild-type from the *t*/+ male to its descendants. The five loci that are required for the TRD consist of one *t* complex responder (*Tcr*) and four *trans*-acting *t* complex distorters (*Tcd*). If a male has a complete set of *Tcd* alleles of *t* haplotype, the chromosome 17 that carries *Tcr* of *t* haplotype is transmitted at high frequency (>95%). These five loci for the TRD have been mapped in the *t* complex (Lyon, 1984) and have been kept as a complete set of TRD alleles of *t* haplotype by the recombination suppression with four large inversions of chromosome (Fig. 1).

Mouse *Tcp-1* is a gene located in the *t* complex. It encodes *t* complex polypeptide 1 (TCP-1) which has homology with "chaperonin" proteins and is thought to have functions in protein folding and assembly (Gupta, 1990). TCP-1 is expressed at a high level during spermatogenesis (Silver *et al.*, 1987a), but its role in TRD is not known. It is also expressed at a lower level in almost all cells investigated. The TCP-1 protein has a *t* haplotype specific form (TCP-1A) and a wild-type specific form (TCP-1B) (Silver *et al.*, 1979; Willison

151

et al., 1986) as detected by 2-dimensional gel electrophoresis. So, we used the gene for evolutionary study of the mouse *t* haplotype.

COMPARISON OF DNA AND AMINO ACID SEQUENCES OF *Tcp-1* BETWEEN WILD-TYPE AND *t* HAPLOTYPE

The *Tcp-1ᵃ* cDNA encoding TCP-1A (Fig. 1) carried by *t* haplotype chromosome was isolated and sequenced from t^{PA027}/t^{PA027} mouse testes (Kubota *et al.*, 1991a). Mouse t^{PA027} haplotype was derived as a rare recombinant from T/t^{w32}(BTBR) mouse and retained *Tcp-1ᵃ* of t^{w32} haplotype chromosome (Morita *et al.*, 1985). The nucleotide sequence of *Tcp-1ᵃ* contained an open reading frame encoding a 556 amino acid polypeptide of 60,400 daltons. On the other hand, the nucleotide sequence of *Tcp-1ᵇ* cDNA encodes a 556 amino acid polypeptide (TCP-1B) with a molecular weight of 60,505 daltons (Willison *et al.*, 1986; Kirchhoff and Willison, 1990).

Kubota *et al.* (1991) described 13 single-bp and one double-bp substitutions between *Tcp-1ᵃ* and *Tcp-1ᵇ* cDNAs in the open reading frame and 2 single-bp substitutions in 5′ and 3′ non-coding regions. Among the substitutions in the coding region, 12 were nonsynonymous and 3 were synonymous, providing the unusual observation that there was a large excess of nonsynonymous substitution. On the contrary, substitutions between mouse *Tcp-1* and the gene encoding human homologue of TCP-1 (Kirchhoff and Willison, 1990) showed normal rates (nonsynonymous/synonymous$=0.17-0.19$). Human and rat comparisons (Morita *et al.*, 1991) also gave a normal rate (0.096). Per-site synonymous and nonsynonymous substitutions gave a high

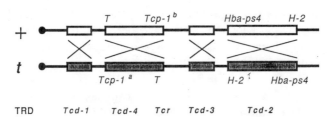

Fig. 1. Genetic maps of a wild-type and a *t*-haplotype mouse chromosome 17. The four inversions are shown by crosses. The genetic symbols *T*, *Tcp-1*, *Hba-ps4*, and *H-2* represent *Brachyury* (short tail), *t complex polypeptide-1*, *hemoglobin alpha-pseudogene 4*, and *histocompatibility-2*, respectively. The loci concerning the transmission ratio distortion are also shown. *Tcd* represents *t* complex distorters and *Tcr*, *t* complex responder. The total length of the map is about 40 Mbp.

nonsynonymous/synonymous rate of substitutions between wild type and *t* haplotype mouse (Kubota *et al.*, 1991).

TCP-1 is a very conserved protein in mammals based on its antigenicity (Willison *et al.*, 1989). In addition, homologues of mouse *Tcp-1* gene have been found in *Drosophila* and yeast (Ursic and Ganetzky, 1988). Thus the *Tcp-1* gene is widespread in eukaryotes. Between mouse and human (Kirchhoff and Willison, 1990), the amino acid substitutions of TCP-1 give values of 2.4–2.6× 10^{-4}/site/Myr and between wild-type mouse and *Drosophila*, 2.4×10^{-4}/site/Myr (Kubota *et al.*, 1991). These similar and rather small values indicate a constant rate of amino acid substitutions in TCP-1 during evolution and give a quantitative support to the idea that TCP-1 is an extremely well conserved protein in eukaryotes. In such cases, amino acid substitutions are well explained by negative selection in the framework of neutral evolution (Kimura, 1983).

However, the diversification of *Tcp-1* gene between *t* haplotype and wild-type is very rapid, because the amino acid substitutions between TCP-1A

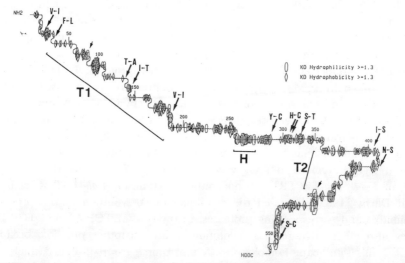

Fig. 2. Schematic secondary structure of mouse TCP-1B predicted by the Chou-Fasman method. Helices predicted are shown with a sine wave, β-sheets with a sharp sawtooth wave, turns by 180 degree directional shifts and coils with a dull sawtooth wave. The domains which would be enriched in β-turns or hydrophilicity are indicated by the characters T1, T2 or H, respectively. The amino acid changes from wild-type (TCP-1B) to *t* haplotype (TCP-1A) by nonsynonymous substitutions are shown by arrows and amino acids. The arrows without the amino acid symbols show the positions of the synonymous substitutions between the two types.

and TCP-1B are 19–42 times greater than those of the above cases (Kubota *et al.*, 1991a). The faster rate of evolution must arise from a high nonsynonymous substitution rate in these two types of mice as described above. Such a high nonsynonymous substitution rate has been reported between the bovine pancreatic trypsin inhibitor and spleen inhibitor genes (Creighton and Charles, 1987). These inhibitors have a very high nonsynonymous/synonymous ratio of substitutions (4.5). Creighton and Charles (1987) suggested it occurs because of strong positive pressure of selection. More recently, Creighton and Darby (1989) indicated, from many sources of data regarding amino acid substitutions between protease inhibitors, that these inhibitors have hypervariability in most functional regions which are involved in physical contact with proteases. These facts give strong support for positive selection occurring in regions of proteins involved in protein-protein interaction. In Fig. 2, we show the predicted secondary structure of TCP-1 protein and the positions of the amino acids substituted between TCP-1A and TCP-1B. T1 and T2 show regions rich in β-turn structure and H is a central hydrophilic region (Chou and Fasman, 1978, 1979). The TCP-1 may have a symmetrical secondary structure which might be involved in oligomeric formation or in its function as a molecular "chaperon". The amino acid substitutions between wild-type and *t* haplotype occur throughout the polypeptide. They may form clusters when they are folded three-dimensionally to interact with other proteins. If TCP-1A and TCP-1B have diverged under positive selection pressure, we propose that TRD might be one mechanism responsible for such selection. TCP-1A and TCP-1B might have rapidly substituted their amino acids in a struggle where TCP-1A increases the transmission ratio of *t* chromosome and TCP-1B increases that of wild-type chromosome. Silver and Remis (1987) suggested that *Tcd-4*, the 4th distorter gene involved in TRD, was located at the *Tcp-1* locus. However, the role of *Tcd-4/Tcp-1* in TRD was not confirmed (Willison *et al.*, 1989). Creighton and Darby (1989) discussed the possibility of co-evolution of the protease inhibitors and corresponding proteases. In the case of TCP-1A and TCP-1B, they also might diverge in co-evolution with some other proteins, because TCP-1 has significant homology to "chaperonin" group proteins which act with smaller subunits in the folding, assembly and transport of other proteins. Such subunits might also be encoded in the *t* complex.

COMPARISON OF *Tcp-1ᵃ* AND *Tcp-1ᵇ* WITH *Tcp-1* HOMOLOGUES OF OTHER SPECIES

  · As *Tcp-1ᵃ* is very similar to human *Tcp-1*, from partial sequence comparison, it has been suggested that the *Tcp-1ᵃ* allele might be the ancestral form of the mouse *Tcp-1* gene (Willison *et al.*, 1987; Delarbre *et al.*, 1988; Ursic and Ganetzky, 1988) and thus the *t* haplotype might be the original type in mice. The comparison of arrangements of *T66* genes in the proximal inversion between *t* haplotype and wild-type has also suggested that the genes in the proximal inversion, including *Tcp-1* gene, were arranged from the *t* haplotype array to the wild-type one (Bullard and Schimenti, 1990). The rat *Tcp-1* cDNA was sequenced and compared (Morita *et al.*, 1991) with TCP-1A and TCP-1B, because rat is more closely related to mouse than man. The mouse *Tcp-1* pseudogene was also sequenced, which originated near the time of the split between mouse and rat and presumably has evolved under no selective pressure. But neither the rat sequence nor the mouse pseudogene sequences have shown any significant difference in their similarity to TCP-1A or TCP-1B (Kubota *et al.*, 1992). So, the ancestral *Tcp-1* gene before the proximal inversion would have no significant difference between its similarity to *Tcp-1ᵃ* and that to *Tcp-1ᵇ* either on the level of nucleotide or encoded amino acid. We call such *Tcp-1* of mouse ancestral type *Tcp-1ᵖ* a prototype. Thus, *Tcp-1ᵃ* and *Tcp-1ᵇ* would have diverged equally from a common ancestor, *Tcp-1ᵖ*, and we cannot say from the sequence analysis which would be the ancestral type.

ORIGIN OF *t* HAPLOTYPES

To investigate the origin of *t* haplotypes and to estimate the accurate time of their spread in *Mus musculus* subspecies, we cloned and partially sequenced the introns of *Tcp-1* structural gene from various mouse species. We made oligonucleotides and amplified genomic DNAs from various mice by the polymerase chain reaction (PCR). Amplified DNA fragments were cloned and sequenced. We analyzed the wild-type *Mus* species and *t* haplotypes from different *M. musculus* subspecies (Morita *et al.*, 1992).

All the four *t* haplotypes investigated so far have an identical *Tcp-1* intron sequence. The *tʷ³²* haplotype is in *M. m. domesticus* and the *tʷ⁷³* haplotype in *M. m. musculus*. The *t* haplotype chromosome of *M. m. molossinus* was from Japan. The three *t* haplotypes among the four are in different complementation groups with regard to lethality factors. The *t* haplotype chromosome was also identified from *M. m. bactrianus*. Thus, the spread of *t* haplotypes extends throughout almost all the *M. musculus* subspecies, but the time of their spread from a unique ancestor is very recent and far later than the split of the *M. musculus* subspecies (1 Myr ago). So, the *t* haplotypes are introgressed to *M. musculus* over the boundaries of subspecies in the world. Then, the polymorphisms among *t* haplotypes, including lethality factors and distal *t* haplotype genetic markers (Hammer *et al.*, 1991) may have been moulded by mutations and recombinations after the recent introgression.

As in the coding region of *Tcp-1* gene, the sequence of *Tcp-1^a* introns carried by the *t* haplotype is very different from that of *Tcp-1^b* carried by other *M. musculus* wild mice. They are calculated to have diverged 2–3 Myr ago from a common ancestor.

From these results and previous findings (Klein *et al.*, 1986; Willison *et al.*, 1986; Silver *et al.*, 1987b; Delarbre *et al.*, 1988; Hammer *et al.*, 1989), we summarize the possible evolution of mouse *t* haplotype (Fig. 3). The proximal inversion of chromosome 17 occurred in an ancestral mouse in the lineage leading to the present-day "wild-type" chromosomes, such as *M. musculus* subspecies (about 2–3 Myr ago). The mice without the proximal inversion would have been ancestors of *t* haplotype mice. Then, during the propagation of mice without proximal inversion the ancestor of *t* haplotype mice would have acquired a high transmission ratio (>95%) in an isolated population. From that time (very recent) the introgression of the *t* haplotype rapidly occurred in *M. musculus* subspecies worldwide by the meiotic drive of TRD. During these processes, the prototype *Tcp-1^p* gene would have evolved to

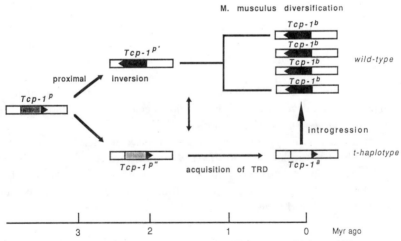

Fig. 3. Postulated events in *t* haplotype evolution. The events which would have occurred on chromosome 17 are summarized. The prototype of *Tcp-1* gene is shown as *Tcp-1^p* and its derivatives as *Tcp-1$^{p'}$* or *Tcp-1$^{p''}$*. Boxes with arrows represent the inverted proximal regions in which *Tcp-1* gene resides.

Tcp-1^b in inversion-bearing animals and to *Tcp-1^a* in mice without inversions, both in an equally rapid manner under some positive pressure.

SUMMARY

Mouse *t* haplotypes are variants of chromosome 17 having four inversions spanning 40 Mbases of DNA. Though the *t* haplotypes have pleiotropic effects on embryonic development, sperm production and fertilization, they spread in the natural population of the house mouse because of higher transmission of *t* haplotype than the wild-type chromosome from heterozygous males. This TRD is one of the cases of meiotic drive. We have cloned and analyzed *Tcp-1* gene, which is in the proximal inverted region of the *t* complex to investigate the evolution of the *t* haplotypes. From the comparison of *Tcp-1* cDNAs between wild and *t* haplotype mice, we observed rapid amino acid substitutions between the two types. This may suggest the positive selection of *Tcp-1* gene divergence involved in TRD. We also sequenced the *Tcp-1* intron 8 and intron 9 from wild mouse species and *t* haplotype mice to estimate the time of their divergence. The comparison showed the origin of the *t* haplotypes to be far earlier than the split of *M. musculus* subspecies. The *Tcp-1* sequences of *t* haplotypes from different *M. musculus* subspecies are almost identical, showing their very recent introgression from the unique ancestral species to *M. musculus* subspecies by TRD.

Acknowledgments

We thank Drs. K. Willison (Institute of Cancer Research, London, U.K.) and N. Takahata (National Institute of Genetics, Mishima, Japan) for valuable discussions. We also thank Drs. G. Gachelin (Pasteur Institute, Paris, France) and K. Moriwaki (National Institute of Genetics, Mishima, Japan) for their kind gifts of mouse DNA. This work was supported by a Grant-in-Aid for Scientific Research (No. 01618508) from the Ministry of Education, Science and Culture of Japan.

REFERENCES

Bullard, D.C. and J.C. Schimenti, 1990. *Genetics* **124**: 957–966.
Chou, P.Y. and G.D. Fasman, 1978. *Adv. Enzymol.* **47**: 45–148.
Chou, P.Y. and G.D. Fasman, 1979. *Biophys. J.* **26**: 367–384.
Committee for Mouse Chromosome 17, 1991. *Mamm. Genome* **1**: 5–29.
Creighton, T.E. and I.G. Charles, 1987. *Cold Spring Harbor Symp. Quant. Biol.* **LII**: 511–519.

Creighton, T.E. and N.J. Darby, 1989. *Trend Biochem. Sci.* **14**: 319-324.

Delarbre, C., Y. Kashi, P. Boursot, J.S. Beckmann, P. Kourilsky, F. Bonhomme, and G. Gachelin, 1988. *Mol. Biol. Evol.* **5**: 120-133.

Frischauf, A.M., 1985. *Trends Genet.* **1**: 100-103.

Gupta, R.S., 1990. *Biochem. Int.* **20**: 833-841.

Hammer, M.F., J. Schimenti, and L.M. Silver, 1989. *Proc. Natl. Acad. Sci. U.S.A.* **86**: 3261-3265.

Hammer, M.G., S. Bliss, and L.M. Silver, 1991. *Genetics* **128**: 799-812.

Kimura, M., 1983. *The Neutral Theory of Molecular Evolution.* Cambridge University Press, Cambridge.

Kirchhoff, C. and K.R. Willison, 1990. *Nucleic Acids Res.* **18**: 4247.

Klein, J., M. Golubic, O. Budimir, T. Schopfer, M. Kasahara, and F. Figueroa, 1986. *Curr. Top. Microbiol. Immunol.* **127**: 239-246.

Klein, J., P. Sipos, and F. Figueroa, 1984. *Genet. Res. Comb.* **44**: 39-46.

Kubota, H., T. Morita, T. Nagata, Y. Takemoto, M. Nozaki, G. Gachelin, and A. Matsushiro, 1991a. *Gene* **105**: 269-273.

Kubota, H., T. Morita, Y. Satta, M. Nozaki, and A. Matsushiro, 1992. *Mamm. Genome* **2**: 246-251.

Lyon, M.F., 1984. *Cell* **37**: 621-628.

Morita, T., C. Delarbre, M. Kress, P. Kourilsky, and G. Gachelin, 1985. *Immunogenetics* **21**: 367-383.

Morita, T., H. Kubota, G. Gachelin, N. Nozaki, and A. Matsushiro, 1991. *Biochim. Biophys. Acta* **1129**: 96-99.

Morita, T., H. Kubota, K. Murata, M. Nozaki, C. Delarbre, K. Willison, Y. Satta, M. Sakaizumi, N. Takahata, and G. Gachelin, 1992. *Proc. Natl. Acad. Sci. U.S.A.* **89**: 6851-6855.

Ruvinsky, A., A. Polyakov, A. Agulnik, T. Herbert, F. Figueroa, and J. Klein, 1991. *Genetics* **127**: 161-168.

Silver, L.M., 1985. *Annu. Rev. Genet.* **19**: 179-208.

Silver, L.M., K. Artzt, and D. Bennett, 1979. *Cell* **17**: 275-284.

Silver, L.M. and D. Remis, 1987. *Genet. Res. Camb.* **49**: 51-56.

Silver, L.M., K.C. Kleen, R.J. Distel, and N.B. Hecht, 1987a. *Dev. Biol.* **119**: 605-608.

Silver, L.M., M. Hammer, H. Fox, J. Garrels, M. Bucan, B. Herrmann, A.-M. Frischauf, H. Lehrach, H. Winking, F. Figueroa, and J. Klein, 1987b. *Mol. Biol. Evol.* **4**: 473-482.

Ursic, D. and B. Ganetzky, 1988. *Gene* **68**: 267-274.

Willison, K.R., K. Dudley, and J. Potter, 1986. *Cell* **44**: 727-738.

Willison, K., A. Kelley, K. Dudley, P. Goodfellow, N. Spurr, V. Groves, P. Gorman, D. Sheer, and J. Trowsdale, 1987. *EMBO J.* **6**: 1867-1974.

Willison, K., V. Lewis, K.S. Zukerman, J. Cordel, C. Dean, K. Miller, M.F. Lyon, and M. Mash, 1989. *Cell* **57**: 621-632.

Evidence for and Implications of Genetic Hitchhiking in the *Drosophila* Genome

CHARLES F. AQUADRO AND DAVID J. BEGUN

Section of Genetics and Development, Biotechnology Building, Cornell University, Ithaca, NY 14853, U.S.A.

The rate of recombination between adjacent base pairs in female *Drosophila melanogaster* has been estimated for several regions of the nuclear genome to be roughly 1×10^{-8} exchanges per generation (Chovnick *et al.*, 1977). In some regions, most notably near the telomere of the X chromosome, near the centromere of the second and third chromosomes, and across the small fourth chromosome of *D. melanogaster*, rates of recombination appear to be substantially lower (*e.g.*, Lindsley and Sandler, 1977). One consequence of the observed rates of recombination is that adjacent nucleotides often have strongly correlated evolutionary histories within populations. For example, the presence of a selectively favored variant at one site can affect the frequency of variants at linked sites by a phenomenon known as the "hitchhiking effect" (Maynard Smith and Haigh, 1974; Ohta and Kimura, 1975; Kaplan *et al.*, 1989).

Consider a population of chromosomes with neutral variation segregating at a large number of sites. Also, for heuristic purposes, imagine that there is no recombination along the chromosome (*e.g.*, animal mitochondrial DNA). If a strongly favored variant arises and is driven to fixation one can see that the entire chromosome containing the selected mutation also goes to fixation. In effect, neutral variation present on all other chromosomes is eliminated from the population. This is an extreme, but instructive example. For most nuclear gene regions, which have at least some recombination, the extent of the

159

reduction of linked neutral variation associated with selective fixations depends on the interplay between the strength of selection, the rate of recombination, and the degree of population subdivision present (Kaplan *et al.*, 1989, 1991; Stephan *et al.*, 1992).

Hitchhiking associated with selective fixation is predicted to have two important consequences for the distribution of DNA variation. The proportion of segregating sites in the vicinity of selected variants is decreased within species or populations. Importantly, the rate of divergence between species at linked neutral sites is not altered by these hitchhiking events (Birky and Walsh, 1988; Hudson, 1990). However, differences between regions in functional constraint and mutation rate (*i.e.*, the neutral mutation rate) will affect both variation within species and divergence between species (Kimura, 1983; Kreitman and Aguadé, 1986). Thus, the comparison of levels of intra- and interspecific variation for different regions of the genome provides a useful approach for distinguishing recent hitchhiking events from strong functional constraint or low mutation rate as the cause of reduced polymorphism in a given region of the genome (Begun and Aquadro, 1991). A second consequence of a recent selective fixation is a skewed frequency distribution for segregating variants, such that rare variants will occur more often than predicted under neutral, equilibrium conditions (Tajima, 1989b; Hudson, 1990).

It is clear that detailed sequencing studies of individual genes has the potential to teach us a great deal about the relative importance of various evolutionary forces determining levels of DNA variation in nature (*e.g.*, Kreitman and Hudson, 1991). However, here we examine levels of variation and divergence over larger segments of chromosomes in *Drosophila* and suggest that such a perspective may add considerably to our understanding of average levels of sequence variation. Specifically, we propose that an important determinant of the level of DNA sequence variation is the physical location of the gene region in question and that this phenomenon is a consequence of the manner in which rates of recombination vary across the *Drosophila* genome. A brief description of these results has appeared elsewhere (Begun and Aquadro, 1992). This hypothesis, if true, suggests that while relatively few nucleotides may be selected at any given time, levels of DNA sequence variation over large sections of *Drosophila* chromosomes may be significantly reduced by genetic hitchhiking associated with selective fixations. Efforts are underway in our lab to obtain the additional data needed to adequately test this hypothesis.

RATES OF RECOMBINATION VARY ACROSS THE *DROSOPHILA* GENOME

The development of detailed polytene chromosome band maps several decades ago provided geneticists with a remarkably detailed physical map of the euchromatic genome of *D. melanogaster*. The plethora of induced and spontaneous mutations available in this species has also provided a rich source

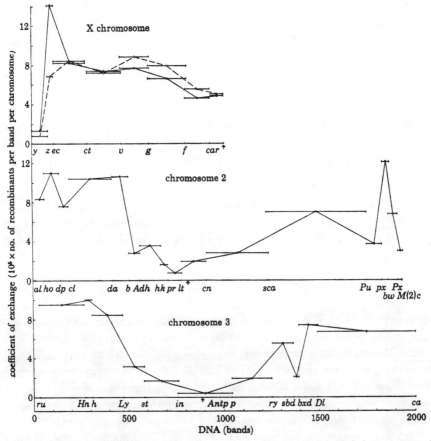

Fig. 1.   Distribution of coefficients of exchange for the X, second and third chromosomes of *D. melanogaster*. The abscissae represent the amount of DNA estimated by the number of polytene bands (solid line). For the X chromosome, the coefficient of exchange is also plotted with the abscissa representing the amount of DNA estimated as a percent of the total DNA content of the euchromatic portion of the X chromosome (from Lindsley and Sandler, 1977).

of material with which to determine a detailed genetic map. The existence of the two types of maps has afforded an unparalleled opportunity to examine the correlation of physical and genetic location. Numerous workers have been interested in the distribution of recombination across the *Drosophila* genome. Morton *et al.* (1976) and Lindsley and Sandler (1977) have summarized available genetic and cytogenetic data which demonstrate that genetic exchange is strongly nonuniform across the euchromatic genome. In particular, the distal tip region of the X chromosome and the euchromatin flanking the centromeres of both the second and third chromosome have extremely low rates of recombination per unit physical length (Fig. 1). The rates of recombination vary in females from approximately 1×10^{-8} exchanges per generation between adjacent nucleotides to levels two or more orders of magnitude lower. On the small fourth chromosome, recombination is normally absent. The reason for this variation in rate of recombination remains unclear, but is associated in some way with proximity to certain telomeric or centromeric sequences (see Lindsley and Sandler, 1977; Ashburner, 1989).

ARE LEVELS OF DNA VARIATION IN *DROSOPHILA* CORRELATED WITH RATE OF RECOMBINATION?

The last decade has brought forth an increasing number of estimates of nuclear DNA sequence variability for random samples of genes from natural populations of *D. melanogaster* and, to a lesser extent, *D. simulans*, *D. pseudoobscura*, and *D. ananassae* (see Aquadro, 1992 for recent review). Table I summarizes levels of DNA variation at 20 gene regions in *D. melanogaster*, together with genetic map position, coefficient of exchange (an estimate of recombination per physical distance) and interspecific divergence (between *D. melanogaster* and *D. simulans*) where available. The data come, with one exception (*cubitus interruptus*), from restriction map surveys of regions 13-65 kb in size, and thus include both coding and noncoding as well as nontranscribed sequences. The estimates of variability and divergence thus provide an average over these different types of regions. The observed heterogeneity in levels of variation between genes in different regions of the genome may be due to differences in functional constraint and/or mutation, but may also reflect the influence of balancing selection and/or genetic hitchhiking associated with selective fixations that have different magnitudes of effect depending on the rate of recombination in the given gene region.

The possibility that levels of variation are reduced in regions of reduced recombination in *Drosophila* has been discussed in the literature for several

TABLE I

Levels of Recombination and DNA Sequence Variation within *D. melanogaster* and Divergence with *D. simulans*

Chromosome/gene region	Location		Coeff. of exchange	Seq. variation and divergence			Ref.
	Genetic	Cytological		$3N\mu/4N\mu$	π	Div.	
X chromosome							
Yellow-achaete (*y, ac*)	0.0	1B1-3	0.0045	0.001	0.001	0.054	1
Phospho-gluconate dehydrogenase (*Pgd*)	0.6	2D4-6	0.0154	0.002	0.003	0.029	1
Zeste-tko (*z, tko*)	1.0	3A3	0.0222	0.004	0.004	—	2
Period (*per*)	1.2	3B1-2	0.0520	0.003	0.001	0.050	1
White (*w*)	1.5	3C2	0.1400	0.007	0.009	—	3
Notch (*N*)	3.0	3C7	0.1212	0.007	0.005	—	4
Vermilion (*v*)	33.0	10A1-2	0.0590	0.004	0.001	0.047	5
Forked (*f*)	56.7	15F1-2	0.0455	0.004	0.002	—	6
Glucose-6-phosphate dehydrogenase (*Zw*)	62.9	18E	0.0485	0.004	0.001	—	7
Suppressor of forked (*su(f)*)	65.9	20F	0.0050	0.000	0.000	—	6
Chromosome II							
sn-glycerol 3-phosphate dehyd. (*Gpdh*)	17.8	25F5	0.0800	0.007	0.008	—	8
Alcohol dehydrogenase (*Adh*)	50.1	35B3	0.0647	0.007	0.006	0.047	9
Dopa decarboxylase (*Ddc*)	53.9	37C1-2	0.0184	0.004	0.005	—	10
Amylase (*Amy*)	77.7	54B1	0.0435	0.006	0.008	—	11
Punch (*Pu*)	97.0	57C5	0.0718	0.004	0.004	—	8
Chromosome III							
Esterase-6 (*Est-6*)	35.9	69A1	0.0604	0.010	0.005	—	12
Metallothionien-A (*MtnA*)	[48]	85E10-15	0.0083	0.002	0.001	0.072	13
Heat shock protein-70A (*Hsp70A*)	[51]	87A6-7	0.0069	0.002	0.002	0.023	14
Rosy (*ry*)	52.0	87D6-13	0.0471	0.004	0.003	0.050	15

Chromosome IV							
Cubitus interruptus Dominant (ci^D)	0.0	102A3	0*	0.000	0.000	0.052	16

Genetic and cytological locations are from Ashburner (1991) and Lindsley and Zimm (1985, 1990). Genetic locations given in brackets are only estimated from the cytological position. Coefficient of exchange was calculated as described in the text (see also Lindsley and Sandler, 1977). Sequence variability was estimated by Hudson's (1982) measure, which estimates $3N\mu$ for X-linked genes and $4N\mu$ for autosomal genes, and by Nei and Li's (1979) measure of nucleotide diversity, π. Divergence was estimated following Nei and Li (1979), and is presented as total divergence, uncorrected for intraspecific polymorphism.
* The recombination rate on the fourth chromosome is effectively zero (see Berry et al., 1991). References: 1) Begun and Aquadro (1991); 2) Aguadé et al. (1988); 3) Miyashita and Langley (1988); 4) Schaeffer et al. (1988); 5) Begun and Aquadro, unpublished results; 6) Langley and Miyashita, unpublished results, cited in Langley (1990); 7) Eanes et al. (1989a); 8) Takano et al. (1991); 9) Langley et al. (1982) and Aquadro et al. (1986); 10) Aquadro, Jennings, Bland, Laurie, and Langley, unpublished results; 11) Langley et al. (1988); 12) Game and Oakeshott (1990), six-cutter data only; 13) Lange et al. (1990); 14) Leigh Brown and Ish-Horowicz (1981) and Leigh Brown (1983); 15) Aquadro et al. (1988); 16) Berry et al. (1991).

years (Aguadé et al., 1989; Beech and Leigh Brown, 1989; Eanes et al., 1989b; Stephan and Langley, 1989; Macpherson et al., 1990; Miyashita, 1990; Langley, 1990). However, only recently was it demonstrated that reduced levels of variation in yellow-achaete (y-ac), a region of reduced recombination near the telomere of the X chromosome in D. melanogaster and D. simulans, result from the hitchhiking effect (Begun and Aquadro, 1991). We showed that D. simulans has dramatically reduced variation at y-ac compared to other gene regions surveyed in this species, and that this reduction could not be attributed to a low neutral mutation rate. The data for D. melanogaster were not as compelling, though there was evidence that variation at y-ac in this species had also been reduced by hitchhiking in at least some populations (Table I; Begun and Aquadro, 1991). Berry et al. (1991) have made a similar observation for the Cubitus interruptus dominant gene region in the tiny fourth chromosome which does not normally undergo recombination.

D. melanogaster and D. simulans last shared a common ancestor 2–3.5 million years ago (Stephens and Nei, 1985). Both species show evidence of selective sweeps in at least some populations. Thus, there have been at least two selective fixations in or near the 35 kb y-ac region over 4–7 million years. With the present data it is not possible to make a more refined estimate. Unfortunately, the fixation of a relatively large region means these data also do not let us distinguish which nucleotide position was the target of selection and which sites were simply swept clean of polymorphism by hitchhiking. This difficulty applies primarily to regions of low recombination or variability. With higher rates of recombination and sufficient density of segregating sites, the focus of selective fixations may be evident as a window of reduced variation centered around the site under selection, analogous to the window of excess polymorphism centered around a putative site of balanced polymorphism at *Adh* (Hudson and Kaplan, 1988; Kreitman and Hudson, 1991). Clearly, the low density of segregating sites in many portions of the *D. melanogaster* genome will make it very difficult to statistically document a reduction in variation. Here, species like *D. simulans* and *D. pseudoobscura*, which have a higher average level of nucleotide polymorphism (*e.g.*, Aquadro *et al.*, 1988; Schaeffer *et al.*, 1987; Riley *et al.*, 1989), will provide greater power to detect selection.

While the fact that hitchhiking may occur in regions of extremely low recombination rates is interesting, a more important issue is the extent of hitchhiking across the entire *Drosophila* genome. What proportion of the genome is affected by selective sweeps? If recombination rates are sufficiently low and/or selective fixations sufficiently frequent, then selective sweeps can effectively eliminate linked neutral variation over a sizeable chromosomal region. Theory suggests that even if the proportion of substitutions that are selectively fixed is very small, the effect of hitchhiking on polymorphism can be quite drastic (Kaplan *et al.*, 1989; Stephan *et al.*, 1992). The hitchhiking effect thus might be expected to be most pronounced in regions of low recombination and to decrease as one moves into regions of higher recombination. There are now enough data from *D. melanogaster* to begin to examine the possibility that there is a general relationship between recombination rates and DNA sequence variation.

One way to explore the relationship between levels of DNA sequence variation and recombination rates is to contrast estimates of nucleotide diversity and the coefficient of exchange (see Lindsley and Sandler, 1977). Nucleotide diversity (π) is the average pairwise difference for all pairs of sequences drawn at random from a population, and can be thought of as heterozygosity per nucleotide (Nei and Li, 1979). The coefficient of exchange for a gene region is calculated by selecting two genetically defined loci which flank the region of interest. One then divides the distance in map units between the flanking loci by the number of polytene bands between the loci (Lindsley and Sandler, 1977). We have used for our calculations the genetic data summarized in Lindsley and Zimm (1985, 1990) and Ashburner (1991) and the cytogenetic data of Bridges' revised maps reproduced in Lindsley and Grell (1968). An

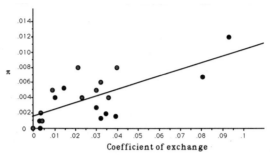

Fig. 2. Scatterplot of nucleotide diversity (π) *vs.* coefficient of exchange for X-linked (closed circles) and autosomal (hatched circles) gene regions in *D. melanogaster*. The values of π for the X chromosome from Table I have been multiplied by four-thirds to make them comparable to estimates of π for autosomal genes due to the difference in effective population size between X-linked and autosomal genes. Coefficients of exchange from Table I have been adjusted to take into account the fact that there is not normally recombination in male *Drosophila* and that X-linked gene regions spend two-thirds of the time in females (where there is recombination) and only one-third of the time in males (where there is no recombination), while an autosome spends half its time in females and half in males. Therefore, we multiplied the coefficient of exchange (from Table I) for autosomal and X-linked gene regions by one-half and two-thirds, respectively, before plotting. Regression line is indicated by a solid line (from Begun and Aquadro, 1992).

important assumption underlying the use of this metric as an index of recombination rate is that over large stretches of the genome (*e.g.*, 20 to 40 polytene bands), the average amount of DNA per polytene band is roughly similar between regions. Available evidence from studies of DNA content of polytene bands suggests that this assumption is reasonable (Lindsley and Sandler, 1977; Sousa, 1988; Merriam *et al.*, 1991).

In Fig. 2 we show a scatterplot of nucleotide diversity (π) *vs.* the coefficient of exchange for gene regions surveyed in *D. melanogaster*. The plotted values represent the estimates of nucleotide variation and coefficients of exchange from Table I corrected for differences between X chromosomes and autosomes in effective population size and rate of recombination (see Fig. 2 legend). There is a general trend for levels of variation to increase as rates of recombination increase. A large fraction of the variation in π is accounted for by variation in the coefficient of exchange, and the null hypothesis that the slope is zero is rejected ($F_s = 16.8$, $p = 0.0007$). Regression requires several assumptions which may not be satisfied by these data. However, the non-parametric Spearman and Kendall tests of rank order correlation are also significantly different from zero (Spearman's $D = 544$, $p < 0.05$; Kendall's tau $= 0.374$, $p < 0.05$). That we can detect a correlation is remarkable given that our estimates of DNA variation and recombination rates are relatively coarse (these errors should not result in any systematic bias) and levels of variation are generally so low in *D. melanogaster*.

There are at least three possible explanations for this apparent trend. First, gene regions in areas of reduced recombination are subject to greater functional constraint (*e.g.*, perhaps the density of transcripts is greater in areas of low recombination). Second, recombination is mutagenic or correlated with higher mutation rates. Third, selective fixations occur relatively frequently throughout the genome with the magnitude of the hitchhiking effect dependent on the recombination rate in a given region.

Ideally we would address the first two possibilities by taking the neutral mutation rate into account when considering levels of variation within species. Differences in neutral mutation rates between gene regions can be detected by comparing levels of divergence between species in these regions. Unfortunately, at this time we only have divergence estimates between *D. melanogaster* and *D. simulans* for a limited number of the gene regions shown in Fig. 2. In Fig. 3 we show a plot of the ratio of nucleotide diversity in *D. melanogaster* to divergence ($\pi/$d) *vs.* coefficient of exchange for X-linked and autosomal gene regions. As in Fig. 2, we have adjusted nucleotide diversity for differences in effective population size for X-linked *vs.* autosomal genes by multiplying the π estimates for autosomal genes in Table I by 0.75 before calculating the ratio to divergence. Note that divergence is also affected by the difference in population size due to the effect of ancestral polymorphism, though the difference relative to the levels of divergence that has accumulated since last sharing a common ancestor appears to be quite small. Coefficients of exchange from Table I are

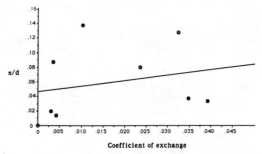

Fig. 3. Plot of $\pi/$d *vs.* coefficient of exchange for eight gene regions in *D. melanogaster* (see text). Filled circles represent X-linked gene regions and hatched circles represent autosomal gene regions. Regression line is indicated by a solid line.

also adjusted before plotting as for Fig. 2. There is a tendency for nucleotide diversity adjusted for interspecific levels of divergence to increase with increasing coefficient of exchange, though it is not statistically significant. This tendency is in the direction expected if variation in mutation rate or functional constraint is not the explanation for the positive correlation between nucleotide diversity and coefficient of exchange seen in Fig. 2. Obtaining divergence estimates for more of the genes in Table I is clearly an important task that will allow more rigorous assessment of the relationship between π/d and coefficient of exchange in *D. melanogaster*.

To explore the possibility that the positive correlation between nucleotide diversity and coefficient of exchange results from recombination *per se* (*i.e.*, that recombination is mutagenic) rather than some population biology phenomenon, we plotted divergence between *D. melanogaster* and *D. simulans*, *versus* the adjusted coefficient of exchange for all gene regions for which we have estimates of divergence (Fig. 4). Even considering that divergence at autosomal genes is expected to be slightly higher than that at X-linked genes due to differences in effective population size and hence ancestral polymorphism, the data at hand provide no support for the hypothesis that gene regions in areas of high recombination rates evolve faster than regions of low rates of recombination. Therefore, at present, we suggest that the positive correlation between DNA variation and rate of recombination shown in Fig. 2 results from an increasing hitchhiking effect associated with selective fixations as recombination rates decrease.

Any correlation between levels of neutral variation and recombination

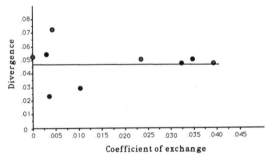

Fig. 4.   Plot of average nucleotide divergence between *D. melanogaster* and *D. simulans*, *versus* the coefficient of exchange for all gene regions for which we have estimates of divergence. Data from Table I. Filled circles represent X-linked gene regions and hatched circles represent autosomal gene regions. Regression line is indicated by a solid line (from Begun and Aquadro, 1992).

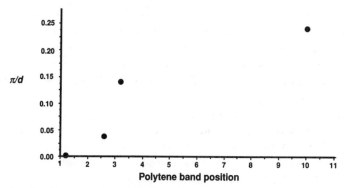

Fig. 5.   Plot of π/d *vs.* cytological band position for four X-linked gene regions in *D. simulans*. Data for *y-ac*, *Pgd*, and *per* are from Begun and Aquadro (1991). The data for vermilion are from Begun and Aquadro (unpublished).

rates may be more apparent in *D. simulans* compared to *D. melanogaster*, as *D. simulans* appears to have higher average levels of sequence variability (reviewed in Aquadro, 1992). In Fig. 5 we show a plot of π/d *vs.* polytene band position (relative to the tip of the X chromosome) for four X-linked gene regions from *D. simulans*. As detailed elsewhere (Begun and Aquadro, 1991), there are not the same empirical estimates of recombination available for *D. simulans* as for *D. melanogaster*. However, the homosequential organization of genes in this region in these two species and observed patterns of linkage disequilibrium for these gene regions in *D. simulans* suggest that rates of recombination are also suppressed near the tip of the X chromosome in *D. simulans* as well as in *D. melanogaster*. There is an obvious increase in variation as one moves away from the tip of the X chromosome. This trend cannot be reasonably attributed to differences in neutral mutation rate because mutation and functional constraint have been factored out by dividing nucleotide diversity by divergence between *D. simulans* and *D. melanogaster*. Hitchhiking effects across the X chromosome may be a sufficient explanation for this pattern in *D. simulans* as well. Given the small amount of data, at the present time we consider this to be an intriguing preliminary observation.

CAN NEUTRAL MODELS EXPLAIN THE CORRELATION?

Can a strict or nearly neutral model account for the observed positive correlation between level of DNA variation and rate of recombination? Results from theory show that, all else being equal, the amount of linked neutral variation is reduced at the time of fixation of a neutral variant, and that the magnitude of the reduction depends on the regional rate of recombination (Tajima, 1990). However, at a random point in time, such as that sampled by an investigator, the average level of neutral DNA diversity is unaffected by the rate of recombination (Hudson, 1983; Tajima, 1990), and the correlation we observe would not be predicted.

Even slightly deleterious mutations have a finite, though extremely small, chance of fixation by genetic drift. Since the conditional time to fixation of a mutant is identical for either positive or negative selection coefficients (see Ewens, 1979), a reduction in flanking neutral variation might also be expected at the time of fixation of these deleterious mutations. However, the number of selective fixations associated with deleterious mutations is expected to be proportionately extremely small. While these types of models need to be considered further in light of our results, they do not seem to be likely explanations at this time.

Thus, the best explanation for the correlation appears to be hitchhiking associated with selective fixations of advantageous mutations. Examination of Fig. 1 reveals that rates of recombination per physical length are reduced not only on the distal segment of the X chromosome and the entire fourth chromosome, but perhaps over as much as 30–40% of each of the second and third chromosomes. Thus, hitchhiking driven by selective fixation of new mutations could play a major role in constraining levels of nucleotide polymorphism over significant portions of the entire *Drosophila* genome. These results provide strong motivation for a detailed analysis of intra- and interspecific variation in more regions across the X-chromosomes and autosomes in both *D. melanogaster* and *D. simulans*.

TWO HITCHHIKING MODELS

There are two sorts of selective hitchhiking explanations for the patterns discussed above. The first is that the genetic composition of the genome is such that regions with reduced recombination rates harbor one or several genes/sequences that are more prone to selective fixations or that have a high likelihood of mutations with meiotic drive. This is a sort of "tail wagging the dog" hypothesis, in which a rapid succession of selective fixations at the tip or centromere has a ripple effect down the chromosome into regions of normal recombination.

A second explanation is that selectively advantageous mutations arise randomly throughout the genome and go to fixation at a roughly constant rate

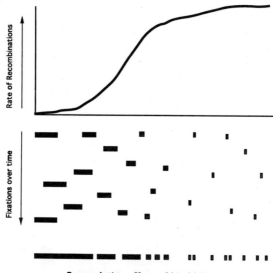

**Cummulative effect of hitchhiking
associated with selective fixations.**

Fig. 6.   Visual summary of the effective size of region affected by hitchhiking associated with selective sweeps in regions of different rates of recombination. The upper plot approximates rates of recombination per physical distance as one moves from the distal tip of the X chromosome towards the centromere, with recombination being extremely low near the tip region and increasing to "normal" levels part way along the X chromosome. The lower plot is a cartoon representing visually (and very roughly) the relative size of regions for which nucleotide diversity might be expected to be reduced by hitchhiking associated with a selective fixation in the approximate middle of each block. The density of shading is meant to indicate the reduction of variation, with black being the most dramatic reduction. Along the bottom of the figure is a rough approximation of the cumulative effect of fixations over some time period in a region with this variation in rate of recombination under this simple model, and implies that levels of nucleotide diversity would be lowest in the region of lowest recombination.

(*i.e.*, average selection coefficients of advantageous mutants are the same across the genome). Provided that a new selectively favored mutation goes to fixation before another advantageous mutation arises in close proximity to it, theory predicts that each fixation will be surrounded by a window of reduced polymorphism, the relative size of which is proportional to the rate of recombination for that region of the genome (Kaplan *et al.*, 1989; Stephan *et al.*, 1992). Hence, near the telomere and centromere, where recombination is very low, each fixation will cause a wide window of reduced polymorphism, while in regions of normal recombination, the window will be proportionately smaller (Fig. 6). As one moves along the chromosome towards regions of lower and lower recombination, the windows become closer and closer, and may begin to overlap substantially. Thus, regions of low recombination are in effect hit by selective sweeps more often, keeping polymorphism at a lower average level. In effect, this model proposes that selective fixations and associated hitchhiking events occur often across the genome, but that our ability to detect them by a reduction in flanking neutral variation is determined by the recombination rate in the region.

RELATIVE FREQUENCY AND CONSEQUENCES OF SELECTIVE FIXATIONS AND ASSOCIATED HITCHHIKING

We have focused our attention here on evidence for reductions of variation that we argue are best explained by hitchhiking associated with selective fixations. There are clearly too few data to draw any conclusion regarding the relative proportion of selective fixations *versus* balanced polymorphisms and simply neutral polymorphisms and random fixations. However, it is worth noting that particularly in regions of low recombination, selective sweeps would appear to have a stronger influence on levels of polymorphism given that they will eliminate or reduce variation that will take on average $4N_e$ generations to regenerate by mutation, where N_e is the effective population size. In contrast, a newly arising balanced polymorphism has its major effect on elevating variation above that expected under neutrality only after a period of accumulation of mutations on the divergent lineages held in the population by selection.

Several points come to mind regarding our ability to determine the proportion of nucleotide sites which have been influenced by selective events. In *D. melanogaster* there may be a bias for detecting balancing selection since selective sweeps are difficult to document statistically at individual gene regions when variation is already relatively low. A great deal of effort is currently being expended in detailed analyses of several gene regions in several species of *Drosophila*. We should keep in mind that if selection on a locus by locus case is very weak (as some quantitative geneticists believe), we may have very little power to detect selection at single genes with currently available statistical tests. The Hudson, Kreitman, and Aguadé (HKA, 1987) test of neutrality is based on the neutrality prediction that the ratio of polymorphism to divergence should be the same throughout the genome. If, however, selective fixations and hitchhiking are pervasive then we may often be comparing two genes or regions, both of which have experienced a reduction in variation as a result of hitchhiking; the HKA test would show no departure from neutrality. This problem can be overcome by the use of reference gene regions for which other types of analysis indicate no evidence of departure from neutrality. Here, the test proposed by Tajima (1989a) has an advantage because it relies only on the distribution of variation in the region under study. However, the power of Tajima's test to detect departures from neutrality due to recent selective fixations is unknown. We feel that the approach outlined here, namely comparing recombination rates and nucleotide diversity across the genome rather than at individual genes, is an alternative which merits further attention.

Hitchhiking may be more prevalent in species with larger population sizes since a larger fraction of advantageous mutants will have their fates determined by selection. Thus, levels of neutral variation would not increase as expected under strict neutrality with an increase in effective population size. This may be another explanation for the relatively narrow range of protein variation observed across taxa with very different census population sizes, as originally proposed by Maynard Smith and Haigh (1974). At this point, however, slightly deleterious selection (*e.g.*, Ohta, 1976; Aquadro *et al.*, 1988; Ohta and Tachida, 1990; Li and Sadler, 1991), population bottlenecks (*e.g.*, Nei and Graur, 1984), and hitchhiking (Kaplan *et al.*, 1989) could all be contributing to the relatively narrow range of protein polymorphism observed in species thought to have radically different population sizes (Lewontin, 1974). Analysis of the distribution of DNA sequence variation within and between species at allozyme-coding genes may provide insight into the relative contributions of these factors.

The depletion of variability by the hitchhiking effect may limit the potential for adaptive evolution. Environments change with time, and variants that are neutral in one situation may be acted upon by selection in another situation or environment. Therefore, potentially advantageous mutations may never have the "opportunity" to be tested by selection because they have been driven from the population by hitchhiking. An intriguing possibility is that this phenomenon could contribute to apparent punctuated patterns of phenotypic evolution revealed in the fossil record. Finally, deleterious mutants will occasionally be driven to fixation by this hitchhiking process, particularly in regions of very low or no recombination (*e.g.*, Rice, 1987).

McDonald and Kreitman (1991) have recently argued that the distribution among three species of *Drosophila* of polymorphic *versus* fixed synonymous and replacement nucleotide variants in the alcohol dehydrogenase gene is incompatible with a neutral model. They suggest that selective fixations of amino acid replacements account for their data, and speculate that as many as

50,000 loci across the genome could be experiencing a selective fixation every 1 to 2 million years. The hypothesis we have suggested in this paper, arrived at by examination of a very different sort of data, similarly proposes that selective fixations may be relatively common. Even so, it is important to note that the selected nucleotides may constitute only a small fraction of the total number of nucleotides going to fixation (Kimura, 1983; Kaplan et al., 1989).

We can make a crude estimate of the number of selective fixations required to predict the data summarized in Fig. 2 as follows. The rate of recombination at the *white* locus is approximately 1×10^{-8} exchanges per nucleotide per generation (B. Judd, cited in Aguadé et al., 1989). From Fig. 2, a 10-fold lower coefficient of exchange corresponds to roughly a 5-fold lower nucleotide diversity. The theoretical results summarized in Fig. 4 of Kaplan et al. (1989) indicate that a reasonable fit to the observed relationship between variation and recombination is obtained with the following parameter values: a mutation rate of 1×10^{-9}, twice the effective population size equal to 1×10^{8}, selective advantage for favored mutations of 1×10^{-2}, rate of recombination varying ten-fold from 1×10^{-8} to 1×10^{-9}, and a ratio of the mutation rate to selectively advantageous mutants compared to the rate of recombination (Λ_r) equal to 2.5×10^{-5}. This leads to the estimate that only 1 in 4,000 fixations is selectively driven, with an average selective advantage of 1%. With this level of selection, the theoretical results of Kaplan et al. (1989) further predict that even the variation in the *white* locus region will be reduced to approximately 40% of the level predicted under strict neutrality at equilibrium.

D. melanogaster and *D. simulans* differ on average by 5% of their nucleotide positions in the euchromatic regions that have been examined. Given that this portion of the genome of both species is approximately 1.15×10^{8} bp in size, then there have been approximately 5.75×10^{6} fixations (due to drift and selection) between these two species in their euchromatin. Since these two species last shared a common ancestor approximately 2.5×10^{6} years ago, then there has been on average one fixation every 1.15 years in each species. As noted above, our rough calculations above assumed that 1 in 4,000 fixations was selectively driven. This would mean that the rate of selective fixation could be as infrequent as once every 3,478 years and still lead to a significant hitchhiking effect and to the correlation we observed. Assuming 10 generations per year, there would be over 30,000 generations between selective fixations. While this is at best a very crude estimate, it is not too dissimilar from that which can be extrapolated from the results of Berry et al. (1991) for *Cubitus interruptus* on the fourth chromosome of *D. melanogaster*. They estimate that there was a 50% chance of a selective sweep on the fourth chromosome in *D.*

melanogaster within the last 2.8×10^5 generations. Given that the fourth chromosome represents approximately one percent of the *Drosophila* coding genome, we can extrapolate from their estimate to a genomic rate of one selective fixation every 2,800 generations. Both of these estimates are well within the maximum rate of selective fixations that a species could sustain estimated by Haldane (1957) from genetic load considerations.

Is a finding that selective sweeps occur with high frequency necessarily incompatible with the observation that many traits in nature are apparently under stabilizing, rather than directional selection? First, we point out that selective fixation may occur so swiftly that the probability of observing the variant being driven to fixation may be quite low. Recall, however, that reduced variation at the DNA level as an aftermath of hitchhiking will be detectable long after the fixation event. Another possibility is that an unusually large number of sweeps have occurred as *D. melanogaster* and *D. simulans* have expanded their ranges into temperate regions in historical times (David and Capy, 1988; Lachaise *et al.*, 1988). This may have resulted in strong new selective pressures at many genes (Singh, 1989). One way to test this idea is to conduct an extensive survey of variation across regions of high and low recombination in a tropical African population. One might predict that levels of variation are higher in these populations and that the correlation between recombination and variation is weaker. However, this prediction is confounded with the influence of effective population size on the magnitude of a selective sweep. Finally, we suggest that to the extent that rates of phenotypic evolution are correlated with the number of selected genes, one might predict that lineages with greater rates of phenotypic evolution will show more evidence of hitchhiking of the type we have discussed here.

CONCLUSIONS

We have examined levels of DNA sequence variation surveyed from natural populations of *D. melanogaster* and *D. simulans* for a number of regions of the genome, exploring particularly evidence for genetic hitchhiking associated with selective fixations. We illustrate the importance of consideration of regional rates of recombination and of levels of divergence (as an index of the neutral mutation rate) in the interpretation of patterns of variation. We want to underscore several points. First is the potential influence of the regional rate of recombination on levels and patterns of DNA sequence variation. Second is the importance of considering the level of divergence (and hence an index of the functional constraint and/or mutation rate) for a given region when interpreting or comparing levels of variation with other regions of the genome. We suggested here that a significant portion of the observed heterogeneity in levels of sequence variation across the *Drosophila* genome may be due to the size of the region affected by adjacent hitchhiking events (which is related to the regional rate of recombination). Clearly, additional heterogeneity in the levels of polymorphism reflects regional differences in the neutral mutation rate as well as the regional influence of balancing selection. We further suggest that studies of DNA variation in *D. simulans* may offer many significant advantages over similar studies in *D. melanogaster* because the higher average level of sequence variability in regions of normal recombination provides more power in statistical tests of neutrality. This is particularly true for detecting selective sweeps associated with the selective fixation of variants. Finally, it will be of particular interest to investigate whether rate of recombination per physical length of region correlates with levels of sequence polymorphism in other organisms, including humans.

SUMMARY

Genetic hitchhiking is a phenomenon whereby the action of natural selection at one nucleotide position influences the amount of neutral variation at linked sites. Hitchhiking associated with the selective fixation of a newly arising advantageous mutation will lead to a reduction in flanking neutral variation. While theory can provide specific predictions regarding the impact of hitchhiking on the distribution of linked neutral variation, it provides no guidance on how often hitchhiking events may occur in nature. We examine levels of DNA sequence variation for 20 gene regions estimated primarily from restriction map surveys throughout the *Drosophila* genome as they relate to empirically determined rates of recombination. We suggest that the level of variation in a gene region is correlated with the rate of recombination in the region, and that this correlation results from hitchhiking effects associated with selective fixations. We suggest that the hitchhiking effect may lead to a reduction of average levels of standing polymorphism in *Drosophila*, not only in regions of severely reduced recombination but throughout a significant portion of the genome.

Acknowledgments

We thank members of our lab and particularly Drs. R.R. Hudson and M. Nachman for fruitful discussion of the ideas presented here, Dr. D. Lindsley for permission to reproduce Fig. 1 from his paper, and Drs. N. Takahata and A.

Clark for constructive criticism of this manuscript. This work was supported by grant GM36431 from the National Institutes of Health, to C.F.A. and a National Institutes of Health predoctoral traineeship to D.J.B.

REFERENCES

Aguadé, M., N. Miyashita, and C.H. Langley, 1988. *Mol. Biol. Evol.* **6**: 123–130.

Aguadé, M., N. Miyashita, and C.H. Langley, 1989. *Genetics* **122**: 607–615.

Aquadro, C.F., 1992. In *Molecular Approaches to Fundamental and Applied Entomology*, edited by J. Oakeshott and M. Whitten. Springer-Verlag, New York (in press).

Aquadro, C.F., S.F. Deese, M.M. Bland, C.H. Langley, and C.C. Laurie-Ahlberg, 1986. *Genetics* **114**: 1165–1190.

Aquadro, C.F., K.M. Lado, and W.A. Noon, 1988. *Genetics* **119**: 875–888.

Ashburner, M., 1989. *Drosophila: A Laboratory Handbook*. Cold Spring Harbor Press, Cold Spring Harbor, New York.

Ashburner, M., 1991. Drosophila Information Service: *Drosophila* Genetic Maps. Vol. 69.

Beech, R.N. and A.J. Leigh Brown, 1989. *Genet. Res.* **53**: 7–15.

Begun, D.J. and C.F. Aquadro, 1991. *Genetics* **129**: 1147–1158.

Begun, D.J. and C.F. Aquadro, 1992. *Nature* **356**: 519–520.

Berry, A.J., J.W. Ajioka, and M. Kreitman, 1991. *Genetics* **129**: 1111–1117.

Birky, C.W. and J.B. Walsh, 1988. *Proc. Natl. Acad. Sci. U.S.A.* **85**: 6414–6418.

Chovnick, A., W. Gelbart, and M. McCarron, 1977. *Cell* **11**: 1–10.

David, J.R. and P. Capy, 1988. *Trends Genet.* **4**: 106–111.

Eanes, W.F., J.W. Ajioka, J. Hey, and C. Wesley, 1989a. *Mol. Biol. Evol.* **6**: 384–397.

Eanes, W.F., J. Labate, and J.W. Ajioka, 1989b. *Mol. Biol. Evol.* **6**: 492–502.

Ewens, W.J., 1979. *Mathematical Population Genetics*. Springer-Verlag, New York.

Game, A.Y. and J.G. Oakeshott, 1990. *Genetics* **126**: 1021–1031.

Haldane, J.B.S., 1957. *J. Genet.* **55**: 511–524.

Hudson, R.R., 1982. *Genetics* **100**: 711–719.

Hudson, R.R., 1983. *Theor. Popul. Biol.* **23**: 183–201.

Hudson, R.R., 1990. In *Oxford Surveys in Evolutionary Biology*, Vol. 7, edited by J. Antonovics and D. Futuyama, pp. 1–44. Oxford University Press, Oxford.

Hudson, R.R. and N.L. Kaplan, 1988. *Genetics* **120**: 831–840.

Hudson, R.R., M. Kreitman, and M. Aguadé, 1987. *Genetics* **116**: 153–159.

Kaplan, N., R.R. Hudson, and M. Iizuka, 1991. *Genet. Res.* **57**: 83–91.

Kaplan, N.L., R.R. Hudson, and C.H. Langley, 1989. *Genetics* **123**: 887–899.

Kimura, M., 1983. *The Neutral Theory of Molecular Evolution*. Cambridge University Press, Cambridge.

Kreitman, M. and M. Aguadé, 1986. *Genetics* **114**: 93–110.

Kreitman, M. and R.R. Hudson, 1991. *Genetics* **127**: 565–582.

Lachaise, D., M.-L. Cariou, J.R. David, F. Lemeunier, L. Tsacas, and M. Ashburner, 1988. *Evol. Biol.* **22**: 159–225.

Lange, B.W., C.H. Langley, and W. Stephan, 1990. *Genetics* **126**: 921–932.

Langley, C.H., 1990. In *Population Biology of Genes and Molecules*, edited by N. Takahata and

J.F. Crow, pp. 75-91. Baifukan, Tokyo.

Langley, C.H., E. Montgomery, and W.F. Quattlebaum, 1982. *Proc. Natl. Acad. Sci. U.S.A.* **79**: 5631-5635.

Langley, C.H., A.E. Shrimpton, T. Yamazaki, N. Miyashita, Y. Matsuo, and C.F. Aquadro, 1988. *Genetics* **119**: 619-629.

Leigh Brown, A.J., 1983. *Proc. Natl. Acad. Sci. U.S.A.* **80**: 5350-5354.

Leigh Brown, A.J. and D. Ish-Horowicz, 1981. *Nature* **290**: 677-682.

Lewontin, R.C., 1974. *The Genetic Basis of Evolutionary Change.* Columbia University Press, New York.

Li, W.-H. and L.A. Sadler, 1991. *Genetics* **129**: 513-523.

Lindsley, D.L. and E.H. Grell, 1968. *Genetic Variations of Drosophila melanogaster.* Carnegie Institution of Washington Publication No. 627.

Lindsley, D.L. and L. Sandler, 1977. *Phil. Trans. R. Soc. Lond. B.* **277**: 295-312.

Lindsley, D. and G. Zimm, 1985. Drosophila Information Service: The Genome of *Drosophila melanogaster.* Vol. 62.

Lindsley, D. and G. Zimm, 1990. Drosophila Information Service: The Genome of *Drosophila melanogaster.* Vol. 68.

Macpherson, J.N., B.S. Weir, and A.J. Leigh Brown, 1990. *Genetics* **126**: 121-129.

Maynard Smith, J. and J. Haigh, 1974. *Genet. Res.* **23**: 23-35.

McDonald, J.H. and M. Kreitman, 1991. *Nature* **351**: 652-654.

Merriam, J., M. Ashburner, D.L. Hartl, and F.C. Kafatos, 1991. *Science* **254**: 221-225.

Miyashita, N.T., 1990. *Genetics* **125**: 407-419.

Miyashita, N. and C.H. Langley, 1988. *Genetics* **120**: 199-212.

Morton, N.E., D.C. Rao, and S. Yee, 1976. *Heredity* **37**: 405-411.

Nei, M. and D. Graur, 1984. *Evol. Biol.* **17**: 73-118.

Nei, M. and W.-H. Li, 1979. *Proc. Natl. Acad. Sci. U.S.A.* **76**: 5269-5273.

Ohta, T., 1976. *Theor. Popul. Biol.* **10**: 254-275.

Ohta, T. and M. Kimura, 1975. *Genet. Res.* **25**: 313-326.

Ohta, T. and H. Tachida, 1990. *Genetics* **126**: 219-229.

Rice, W.R., 1987. *Genetics* **116**: 161-168.

Riley, M.A., M.E. Hallas, and R.C. Lewontin, 1989. *Genetics* **123**: 359-369.

Schaeffer, S.W., C.F. Aquadro, and W.W. Anderson, 1987. *Mol. Biol. Evol.* **4**: 254-265.

Schaeffer, S.W., C.F. Aquadro, and C.H. Langley, 1988. *Mol. Biol. Evol.* **5**: 30-40.

Singh, R.S., 1989. *Annu. Rev. Genet.* **23**: 425-453.

Sousa, V., 1988. *Chromosome Maps of Drosophila.* CRC Press, Inc., Boca Raton, Florida.

Stephan, W. and C.H. Langley, 1989. *Genetics* **121**: 89-99.

Stephan, W., T.H. E. Wiehe, and M.W. Lenz, 1992. *Theor. Popul. Biol.* **41**: 237-254.

Stephens, J.C. and M. Nei, 1985. *J. Mol. Evol.* **22**: 289-300.

Tajima, F., 1989a. *Genetics* **123**: 585-595.

Tajima, F., 1989b. *Genetics* **123**: 597-601.

Tajima, F., 1990. *Genetics* **125**: 447-454.

Takano, T.S., S. Kusakabe, and T. Mukai, 1991. *Genetics* **129**: 753-761.

DNA Polymorphism and the Origin of Protein Polymorphism at the *Gpdh* Locus of *Drosophila melanogaster*

TOSHIYUKI S. TAKANO, SHINICHI KUSAKABE,[*1] AND
TERUMI MUKAI[*2]

Department of Biology, Kyushu University, Fukuoka 812, Japan

Classical population genetics has revealed that there is an abundance of protein polymorphism in natural populations. *Drosophila* is one of the best-studied organisms in this respect. While most population geneticists agree that protein polymorphism is primarily maintained by mutation and random genetic drift (Kimura and Ohta, 1971; Kimura, 1983), it is also known that electrophoretic variants at several enzyme loci exhibit latitudinal clines. The most well-studied examples in *D. melanogaster* are the fast/slow polymorphisms at the alcohol dehydrogenase (*Adh*) and the *sn*-glycerol-3-phosphate dehydrogenase (*Gpdh*) locus (Oakeshott *et al.*, 1982). Similarly, analyses of inversion frequencies have shown the existence of latitudinal clines for frequencies of four cosmopolitan inversions, *In(2L)t*, *In(2R)NS*, *In(3L)P*, and *In(3R)P* (Mettler *et al.*, 1977; Knibb, 1982). Although strong linkage disequilibria between allozyme genes and their linked inversions have been identified, disequilibrium with inversion *In(2L)t* is only a partial explanation of the clines of the *Adh* and *Gpdh* genes (Voelker *et al.*, 1978). These major polymorphisms and clines suggest that some balancing selection has operated in natural populations.

Since Kreitman (1983) reported 11 *Adh* sequences of *D. melanogaster*, data

---

[*1] Present address: Department of Biology, Faculty of Integrated Arts and Sciences, Hiroshima University, Hiroshima 730, Japan.

[*2] Died April 19, 1990.

on intraspecific DNA polymorphism and interspecific DNA divergence at the *Adh* locus have been accumulated from both restriction map survey and direct DNA sequencing (*e.g.*, Bodmer and Ashburner, 1984; Aquadro *et al.*, 1986; Kreitman and Aguadé, 1986a; McDonald and Kreitman, 1991). One of the most remarkable features obtained is an excess of intraspecific silent polymorphism in the coding region in comparison with a 5′ flanking region and introns, whereas the amount of interspecific DNA variation in the exons approximately equals that of the 5′ flanking region (Kreitman and Aguadé, 1986b; Hudson *et al.*, 1987). In addition, the origin of fast/slow polymorphism at the *Adh* locus is found to be very old (Stephens and Nei, 1985).

On the other hand, relatively little is known about the patterns of DNA polymorphism at the *Gpdh* locus. There are three distinct isozymes, designated as GPDH-1, -2, and -3, which exhibit a unique temporal and tissue-specific pattern of expression. This gene has been well characterized at the level of fine structure and is composed of eight exons with three classes of transcripts produced by alternative processing of 3′-end exons (von Kalm *et al.*, 1989; Bewley *et al.*, 1989).

The present paper reports DNA divergence and polymorphism at the *Gpdh* locus among one *D. simulans* and 11 *D. melanogaster* alleles sampled from six distinct geographic locations. In addition to the latitudinal clines, the observed patterns of DNA polymorphism at the *Gpdh* locus were very similar to those of the *Adh* locus.

DROSOPHILA STOCKS

An inbred line of *D. simulans*, which originated from an isofemale line collected in Japan, was established by brother-sister mating for 20 generations. Six lines of *D. melanogaster* were sampled from three populations: Raleigh, NC (abbreviated RA); Chichijima Island, Ogasawara (OG); and Aomori, Japan (AO). The constructions of lines from the Raleigh population were described in Koga *et al.* (1991), and those from the Aomori and Ogasawara populations in Takano *et al.* (1991). Each chromosome line was characterized as to α-GPDH allozyme phenotype and presence of second chromosome inversions. Two chromosomes bearing the fast allozyme and standard arrangement (RA36-F and OG278-F), two slow and standard chromosomes (OG315-S and AO83-S), and two fast and *In(2L)t*-carrying chromosomes (RA79-I and AO194-I) were sampled. In addition, a few second chromosomes were extracted in the same manner from three geographically distinct populations: Tananarive, Madagascar (J); Nairobi, Kenya (L); Pakistan (PAK). Thus, two

fast (J72-F and L90-F) and three slow (J93-S, L149-S, and PAK2-S) chromosomes were utilized in the second experiment.

CLONING AND SEQUENCING

Genomic libraries were constructed in λCharon 35 or λDASHII with
*Sau*3AI (*Mbo*I) or *Hind*III for a strain of *D. simulans*, RA36-F, RA79-I, and
L149-S. Partial libraries from complete digestion by *Sal*I or *Eco*RI were made
in λDASHII for the remaining eight strains. Genomic clones of the *Gpdh* genes
were obtained by standard procedures using *pDm60a(c)* (Cook *et al.*, 1986) and
in other cases *pG8Sl* and *pG9E* (Takano *et al.*, 1989) as probes. Each *Gpdh* clone
was subcloned into pUC13 or Bluescript M13. Both strands of each gene were
sequenced by the dideoxyribonucleotide chain termination method (Sanger *et
al.*, 1977) and by using 17 base oligonucleotide primers that were spaced at
approximately 300 bp intervals. An allele of *D. simulans* and six *D. melanogaster*
alleles (RA36-F, OG278-F, OG315-S, AO83-S, RA79-I, and AO194-I) were
sequenced for at least 5,286 bp from a *Hind*III site at position 0 to the first

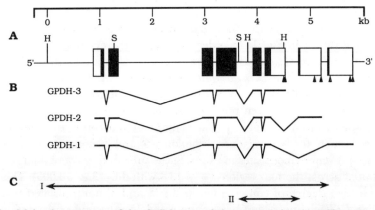

Fig. 1.   Molecular structure of the *Gpdh* locus and the sequenced regions. Fine structure of
the *Gpdh* locus is shown in (A) (Bewley *et al.*, 1989), where shaded and open boxes indicate
coding and nontranslated regions, respectively. The positions of potential polyadenylation
signals are shown by arrowheads. H and S refer to the restriction endonucleases *Hind*III and
*Sac*I. (B) represents three transcripts from the *Gpdh* locus, designated as GPDH-1, -2, and -3.
Bewley *et al.* (1989) have reported only the 3′-end of GPDH-2, but a complete structure of
GPDH-2 has been illustrated following their suggestion. The sequenced regions are shown in
(C). I is the region sequenced for a *D. simulans* and six *D. melanogaster* strains, and five other
D. melanogaster strains were sequenced only for region II (see text).

TABLE I

Distribution of Nucleotide Differences between *D. melanogaster* (RA36-F) and *D. simulans*

	5′ flanking	*Gpdh* locus		
		Nontranslated	Introns	Exons
No. of silent sites compared	859	875	2,454	251
No. of nucleotide differences	59	26	120	15
No. of nucleotide substitutions per site ($\pm$SE)	0.0720 ± 0.0095	0.0303 ± 0.0060	0.0506 ± 0.0047	0.0623 ± 0.0163
		0.0513 ± 0.0035		

The numbers of substitutions per site and their standard errors were estimated as described in Jukes and Cantor (1969) and Kimura and Ohta (1972), respectively.

potential polyadenylation signal of exon 8, and the other five alleles were sequenced from a *Sac*I site in the 4th intron to the end of exon 7 (see Fig. 1).

DNA DIVERGENCE BETWEEN *D. MELANOGASTER* AND *D. SIMULANS*

The organization of the *Gpdh* gene in *D. melanogaster* and the sequenced regions are shown in Fig. 1. In order to estimate the evolutionary rate at the *Gpdh* gene, a *D. simulans Gpdh* gene was sequenced for a total of 5,339 bp and aligned with one of the *D. melanogaster* sequences (RA36-F), which bears the fast allozyme for GPDH. This study revealed 220 nucleotide substitutions and 32 length variations between the two sequences. A summary of the sequence comparison is given in Table I. Calculations of nucleotide divergence were made separately for four regions: 5′ flanking, 5′ and 3′ nontranslated regions, introns, and synonymous sites in exons. Silent substitutions were nonrandomly distributed across the four regions ($\chi^2 = 13.8$ with d.f. $= 3$, $p < 0.005$). This may reflect a high degree of conservation in the 5′ and 3′ nontranslated regions.

The *Gpdh* coding region has accumulated 15 nucleotide substitutions, all of which occurred in synonymous sites. As mentioned above, there are two electrophoretic variants (fast and slow) at the *Gpdh* locus in *D. melanogaster*. The lack of difference in deduced amino acid sequences between the two species suggests that the fast allozyme is an ancestral allele. Further supporting evidence will be mentioned later. This is also consistent with the finding that the GPDH protein is highly conserved and is evolving slowly (Bewley *et al.*, 1989).

DNA POLYMORPHISM WITHIN *D. MELANOGASTER*

Six *D. melanogaster Gpdh* alleles were sequenced and aligned for a total of 5,347 bp, excluding insertions/deletions. Two of the six alleles carry the fast allozyme for GPDH, two the slow, and the remaining two chromosomes bear *In(2L)t* and the fast allozyme. Only one amino acid replacement was found among 109 segregating sites observed in this study. This change is AAT (Asn) $\rightleftarrows$ AAA (Lys) in exon 6 and is responsible for the two electrophoretic variants (Kusakabe *et al.*, 1990).

Table II shows a summary of DNA sequence variation at silent sites. Surprisingly, the proportion of polymorphic sites at synonymous sites in exons is about 3 times higher than corresponding proportions in the other three regions, a finding very similar to the results observed at the *Adh* gene (Kreitman and Aguadé, 1986b; Hudson *et al.*, 1987). This tendency does not disappear even if only the two fast alleles are compared, in which case the number of nucleotide differences per silent site in exons is 0.056. It should be mentioned here that the present θ values are not an estimate of $4N\mu$ (N is the effective population size and μ the mutation rate per site) in this region since the chromosomes were not randomly sampled. At any rate, the estimate of θ at silent sites in exons is about 5 times as large as the average estimate of Hudson's θ (Hudson, 1982) obtained from a restriction map survey (0.005; *cf.* Aguadé *et al.*, 1989). Indeed, Hudson's θ in a 23-kb region including the *Gpdh* gene is estimated to be 0.0066 from a restriction map survey (Takano *et al.*, 1991). It appears that a high degree of silent polymorphism in the coding region is maintained in natural populations of *D. melanogaster*.

TABLE II
Distribution of Silent Site Variation among Six *D. melanogaster* Sequences

	5′ flanking	*Gpdh* locus		
		Nontranslated	Introns	Exons
No. of silent sites compared	883	875	2,491	251
No. of segregating sites	20	18	54	16
$\hat{\theta}^a$	0.0099	0.0090	0.0095	0.0279

[a] $\hat{\theta}$ was estimated by $\dfrac{p}{\sum_{i=1}^{n-1}\frac{1}{i}}$, where p is the proportion of segregating sites in the sample and n is a sample size.

The numbers of transitions and transversions were approximately equal to each other, 53 transitions and 56 transversions including an amino acid replacement. This observed ratio of transitions : transversions is significantly different from the expected ratio of 1 : 2 ($\chi^2 = 11.5$ with d.f. $= 1$, $p < 0.001$). This ratio for polymorphism in exons was 10 : 7, which again differs significantly from the expected ratio ($\chi^2 = 4.97$ with d.f. $= 1$, $p < 0.05$). Two types of transitions occurred equally: the numbers of A$\rightleftarrows$G and C$\rightleftarrows$T transitions were 26 and 27, respectively.

ORIGIN OF FAST AND SLOW ELECTROPHORETIC POLYMORPHISM

From the above analysis of six *D. melanogaster* sequences, we can obtain the amount of variation within-fast and within-slow alleles; nucleotide diversities between the two fast and the two slow alleles were 0.0127 and 0.0100, respectively. This does not clearly indicate that the origin of the fast/slow polymorphism is very old, but intragenic recombination or gene conversion appears to occur in this region. A phylogenetic analysis of the region from exon 4 to exon 7 shows the fast-slow allelic dichotomy, whereas clusters of the fast and slow alleles were not observed in a region from the 5′ flanking region to the third intron (data not shown). Thus, in order to exclude effects of recombination and gene conversion as much as possible, only a 1,128 bp region (907.7 silent sites) from the *Sac*I site in 4th intron to exon 7 was utilized in the analysis, where two fast and three slow alleles from three geographically distinct populations were added. The number of nucleotide substitutions per silent site between 11 alleles was estimated as described in Jukes and Cantor (1969), and

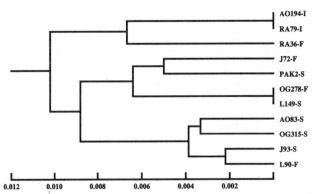

Fig. 2.   UPGMA tree of 11 *D. melanogaster* sequences based on the estimates of the number of nucleotide substitutions per site in 907.7 silent sites.

a UPGMA tree was constructed from these estimates. The result is illustrated in Fig. 2. The fast and slow alleles did not exhibit clustering, which suggests frequent recombination and/or gene conversion in this short region or recurrent mutations at the fast/slow replacement site. If we assume that there was no recombination and no gene conversion, four independent mutations at the fast/slow replacement site were required for the tree in Fig. 2. The most reliable estimate of recombination rate, obtained from the analyses at the *rosy* region, is approximately 10^{-8} between adjacent base pairs (Chovnick *et al.*, 1977). The frequency of occurrence of recombination at the *Gpdh* locus may be still higher than this estimate since the *rosy* locus is located in a region exhibiting a somewhat reduced rate of exchange due to the centromere effect (Gelbart *et al.*, 1976). Gene conversion is also observed with a compatible frequency at the *rosy* locus (Curtis *et al.*, 1989). Some recombination and gene conversion appear to have participated in formation of the present phylogenetic relationship among the alleles.

Since polymorphic duplications at the *Gpdh* locus have been found in natural populations of *D. melanogaster* (Koga *et al.*, 1988; Takano *et al.*, 1989), the situation at the *Gpdh* locus is more complicated. The structure of this duplication has been studied at the DNA sequence level (Kusakabe *et al.*, 1990). This duplication deletes the first and second exons, resulting in a kind of pseudogene, and the duplicated genes appear to evolve in a concerted manner by gene conversion or unequal crossing over. The present sample includes two duplication-bearing chromosomes (L149-S and J93-S). Nucleotide sequences from the duplicated genes of J93-S were identical to each other in the region studied. An accurate estimate of recombination and gene conversion at the *Gpdh* locus is needed to reconstruct a phylogenetic tree of allelic variants and to infer the evolutionary history of the fast and slow alleles.

Although we cannot estimate the age of the electrophoretic polymorphism from the present data, the data suggest that the fast alleles are ancestral to the

TABLE III

Silent Site Variation within Fast and Slow Alleles

	Number of chromosomes	Number of segregating sites	$\hat{\theta}$	Nucleotide diversity
Within fast	4	34	0.0204	0.0202 ± 0.0038
Within slow	5	21	0.0111	0.0117 ± 0.0026

DNA variation was estimated from the data in a 1,128 bp region (907.7 synonymous sites) including the fast/slow replacement site. Nucleotide diversity and its sampling variance were calculated following Nei and Tajima (1981).

slow. One piece of evidence which has been described above is obtained from a comparison of the deduced amino acid sequences. In addition, Lakovaara and Keränen (1980) have reported that 10 out of 12 species of the *D. melanogaster* group studied are monomorphic for the fast electrophoretic allele in *D. melanogaster*, although more recently two electrophoretic variants have been observed in a local population of *D. simulans* (Choudhary and Singh, 1987) which is under the class of monomorphic species according to Lakovaara and Keränen. The second piece of evidence concerns within allele variation. DNA variation among the four fast chromosomes was about 2 times as large as that among the five slow alleles (Table III), suggesting that the slow is a more recent allele.

CONCLUSION

The present analysis of one *D. simulans* and 11 *D. melanogaster* *Gpdh* sequences suggests that the electrophoretic "slow" variant is a more recent allele. Frequencies of the slow allele increase as distance from the equator increases, and a significant correlation of frequency with latitude has been observed in North America (Oakeshott *et al.*, 1982) and in Australasia (Oakeshott *et al.*, 1984). Thus the latitudinal cline cannot be explained by the recent origin and spread of the slow allele. The latitudinal cline for the *Adh* locus is much more pronounced, and the relationship of allozyme frequency with latitude is statistically significant in all three continents studied, North America, Australasia, and Asia-Europe (Oakeshott *et al.*, 1982). Again, a more recent allele, the fast allele at the *Adh* locus, is found in high frequency at higher latitudes in these clines. Frequency distributions of the more recent alleles confirm that simple historical explanations seem inadequate to account for the widespread occurrence of these clines.

The present study reveals an excess of silent polymorphisms in the coding region in contrast to those in a 5′ flanking region and introns. This is also compatible with the finding for the *Adh* locus (Kreitman and Aguadé, 1986b; Hudson *et al.*, 1987; Kreitman and Hudson, 1991). A comparison of silent polymorphism between the *Gpdh* and *Adh* loci is summarized in Fig. 3. The other notable feature of the *Adh* locus is a clustering of polymorphisms in exon 4, which contains the amino acid replacement polymorphism distinguishing the two allozymes. On the other hand, 10 out of 16 silent polymorphisms detected in the *Gpdh* coding region among six *D. melanogaster* sequences were observed in exon 4 (data not shown), whereas the replacement polymorphism was found in exon 6. This is a major difference in the pattern of DNA polymorphisms

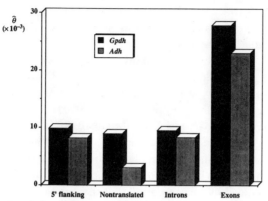

Fig. 3. Comparison of estimates of θ in the *Gpdh* and *Adh* loci. The estimates in the *Adh* locus were obtained from Kreitman and Hudson (1991).

TABLE IV
Distribution of Silent Site Variation among 11 *D. melanogaster* sequences

	Intron	Exon 5	Intron	Exon 6	Nontrans- lated	Intron	Exon 7
Total length	264	154	68	94	281	236	31
Silent sites compared	264	35.3	68	18.3	278.7	236	7.3
Number of segregating sites	16	3	0	1	15	3	0
$\hat{\theta}$	0.0207	0.0290	0.0	0.0186	0.0184	0.0043	0.0

between the two genes. However, it should be noticed here that exon 6 is a very short coding region. A summary of silent site variation in a 1,128 region from the *Sac*I site in the 4th intron to exon 7 is shown in Table IV, where we can see a high degree of polymorphism from intron 4 to the nontranslated region of exon 6, covering 664.3 silent sites. The estimate of θ in this region was 0.018 and this is about twice as large as that in a 5′ flanking region. Thus, exon 6 appears to be included in a region manifesting a high degree of polymorphism.

We cannot estimate the age of the two electrophoretic variants due to the complicated phylogenetic relationship of the DNA sequences. Surprisingly, one of the fast alleles (OG278-F) has exactly the same sequence as one of slow alleles (L149-S), excluding an amino acid replacement change. Curtis and Bender (1991) studied gene conversion at the *rosy* locus and estimated the average length of gene conversion tracts as about 1.2 kb, ranging from 500 bp to greater than 3 kb. Very short gene conversion appears to be unusual. Double crossing over in a short region is also very rare. Recurrent mutation may be responsible for the phenomenon.

The present sample includes two inversion-carrying chromosomes (RA79-I and AO194-I). These alleles were sequenced for 5,389 bp, which corresponds to 4,542 silent sites. Only two sites were different from each other, thus the number of nucleotide differences per site was 0.0004. The origin of *In(2L)t* in a local population in Japan has been estimated from known recombination frequencies and the low level of protein polymorphism among inversion-carrying chromosomes to have taken place roughly a few thousand generations ago (Mukai *et al.*, 1980). These results are consistent with each other. On the other hand, Aguadé (1988) has reported that the estimate of heterozygosity per site (Engels, 1981) was 0.0046 in the Spanish samples, which indicates that the origin of *In(2L)t* is also very old. If this is the case, a few chromosomes carrying *In(2L)t* have migrated to Japan, and probably to North America, and increased rapidly in frequency. This finding is surprising when we consider that a restriction map survey has demonstrated nearly equivalent variation among samples from geographically distinct localities. Indeed, six out of seven segregating restriction sites among chromosomes with the standard arrangement from the Spanish samples were also observed in the North American samples (Kreitman and Aguadé, 1986a). The reasons for the increase in the inversion frequency and the mechanism responsible for the inversion clines are unknown. Further study is necessary to answer these questions.

Kreitman and Hudson (1991) have suggested that a single balanced polymorphism, possibly the fast/slow allozyme polymorphism, is responsible for the pattern of DNA variation observed. We found a relatively high degree of DNA polymorphism around the amino acid replacement site responsible for the fast/slow allozyme polymorphism, but at the same time there was also a complicated phylogenetic relationship of the fast and slow alleles. We should note that even if frequent exchange of DNA occurs between the fast and slow alleles or multiple origins of the slow alleles, there is nevertheless a latitudinal cline of allozyme frequency. The fast/slow polymorphism itself might still have evolutionary significance. Our analysis does not permit identification of the factors primarily responsible for the patterns of DNA polymorphism and the allozyme cline. We were able to demonstrate, however, that the patterns of polymorphism in the *Gpdh* locus were very similar to those found in the *Adh* locus.

SUMMARY

The *Gpdh* gene was isolated from one *D. simulans* and six *D. melanogaster* strains, and the DNA sequence of a region of about 5.3 kb including the structural gene and flanking sequence was determined for each. Five other *D. melanogaster* strains were sequenced for a 1,128 bp region including the segregating site responsible for the fast/slow allozyme. The main findings are as follows: (1) the estimate of the number of substitutions per site between *D. simulans* and *D. melanogaster* was about 0.05. (2) Among six *D. melanogaster* sequences, there is an excess of silent polymorphism in the coding region, in contrast to those in a 5′ flanking region and introns. This pattern was not seen in interspecific comparisons. (3) The within-allozyme variation and a comparison of deduced amino acid sequences between the two species suggest that the electrophoretic "slow" allele is a more recent allele. Accumulated findings suggest that the observed patterns of DNA polymorphism and DNA divergence at the *Gpdh* locus are very similar to those of the *Adh* locus, as are latitudinal clines for the fast/slow polymorphism. Frequency distributions of the more recent alleles (the slow allele at the *Gpdh* locus and the fast at the *Adh* locus) indicate that simple historical explanations are inadequate to account for the widespread occurrence of the latitudinal clines at these loci.

Acknowledgments

We would like to thank Drs. T.K. Watanabe and S.C. Ishiwa for providing strains of *D. simulans* and *D. melanogaster* isofemale lines collected in Africa and Pakistan. We also thank Drs. A. Clark, N. Takahata, and S. Pugh for improving the manuscript. This investigation was supported by research grants from the Ministry of Education, Science and Culture of Japan.

REFERENCES

Aguadé, M., 1988. *Genetics* **119**: 135–140.
Aguadé, M., N. Miyashita, and C.H. Langley, 1989. *Genetics* **122**: 607–615.
Aquadro, C.F., S.F. Desse, M.M. Bland, C.H. Langley, and C.C. Laurie-Ahlberg, 1986. *Genetics* **114**: 1165–1190.
Bewley, G.C., J.L. Cook, S. Kusakabe, T. Mukai, D.L. Rigby, and G.K. Chambers, 1989. *Nucleic Acids Res.* **17**: 8553–8567.
Bodmer, M. and M. Ashburner, 1984. *Nature* **309**: 425–430.
Choudhary, M. and R.S. Singh, 1987. *Genetics* **117**: 697–710.
Chovnick, A., W. Gelbart, and M. McCarron, 1977. *Cell* **11**: 1–10.

Cook, J.L., J.B. Shaffer, G.C. Bewley, R.J. MacIntyre, and D.A. Wright, 1986. *J. Biol. Chem.* **261**: 11751-11755.

Curtis, D. and W. Bender, 1991. *Genetics* **127**: 739-746.

Curtis, D., S.H. Clark, A. Chovnick, and W. Bender, 1989. *Genetics* **122**: 653-661.

Engels, W.R., 1981. *Proc. Natl. Acad. Sci. U.S.A.* **78**: 6329-6333.

Gelbart, W., M. McCarron, and A. Chovnick, 1976. *Genetics* **84**: 211-232.

Hudson, R.R., 1982. *Genetics* **100**: 711-719.

Hudson, R.R., M. Kreitman, and M. Aguadé, 1987. *Genetics* **116**: 153-159.

Jukes T.H. and C.R. Cantor, 1969. In *Mammalian Protein Metabolism III*, edited by H.N. Munro, pp. 21-132. Academic Press, New York.

Kimura, M., 1983. *The Neutral Theory of Molecular Evolution*. Cambridge University Press, Cambridge.

Kimura, M. and T. Ohta, 1971. *Nature* **229**: 467-469.

Kimura, M. and T. Ohta, 1972. *J. Mol. Evol.* **2**: 87-90.

Knibb, W.R., 1982. *Genetica* **58**: 213-221.

Koga, A., S. Kusakabe, F. Tajima, K. Harada, G.C. Bewley, and T. Mukai, 1988. *Proc. Jpn. Acad.* **64, Ser. B**: 9-12.

Koga, A., S. Kusakabe, F. Tajima, T. Takano, K. Harada, and T. Mukai, 1991. *Genet. Res.* **58**: 145-156.

Kreitman, M., 1983. *Nature* **304**: 412-417.

Kreitman, M. and M. Aguadé, 1986a. *Proc. Natl. Acad. Sci. U.S.A.* **83**: 3562-3566.

Kreitman, M.E. and M. Aguadé, 1986b. *Genetics* **114**: 93-110.

Kreitman, M. and R.R. Hudson, 1991. *Genetics* **127**: 565-582.

Kusakabe, S., H. Baba, A. Koga, G.C. Bewley, and T. Mukai, 1990. *Proc. R. Soc. Lond. B.* **242**: 157-162.

Lakovaara, S. and L. Keränen, 1980. *Hereditas* **92**: 251-258.

McDonald, J.H. and M. Kreitman, 1991. *Nature* **351**: 652-654.

Mettler, L.E., R.A. Voelker, and T. Mukai, 1977. *Genetics* **87**: 169-176.

Mukai, T., H. Tachida, and M. Ichinose, 1980. *Proc. Natl. Acad. Sci. U.S.A.* **77**: 4857-4860.

Nei, M. and F. Tajima, 1981. *Genetics* **97**: 145-163.

Oakeshott, J.G., J.B. Gibson, P.R. Anderson, W.R. Knibb, D.G. Anderson, and G.K. Chambers, 1982. *Evolution* **36**: 86-96.

Oakeshott, J.G., S.W. McKechnie, and G.K. Chambers, 1984. *Genetica* **63**: 21-29.

Sanger, F., S. Nicklen, and A.R. Coulson, 1977. *Proc. Natl. Acad. Sci. U.S.A.* **74**: 5463-5467.

Stephens, J.C. and M. Nei, 1985. *J. Mol. Evol.* **22**: 289-300.

Takano, T., S. Kusakabe, A. Koga, and T. Mukai, 1989. *Proc. Natl. Acad. Sci. U.S.A.* **86**: 5000-5004.

Takano, T.S., S. Kusakabe, and T. Mukai, 1991. *Genetics* **129**: 753-761.

Voelker, R.A., C.C. Cockerham, F.M. Johnson, H.E. Schaffer, T. Mukai, and L.E. Mettler, 1978. *Genetics* **88**: 515-527.

von Kalm, L., J. Weaver, J. DeMarco, R.J. MacIntyre, and D.T. Sullivan, 1989. *Proc. Natl. Acad. Sci. U.S.A.* **86**: 5020-5024.

Molecular Genetic Studies of Postmating Reproductive Isolation in *Drosophila*

CHUNG-I WU, DANIEL E. PEREZ, ANDREW W. DAVIS,
NORMAN A. JOHNSON, ERIC L. CABOT,
MICHAEL F. PALOPOLI, AND MAO-LIEN WU

Department of Ecology and Evolution, University of Chicago, Chicago, IL 60637, U.S.A.

The genetic basis of reproductive isolation is undoubtedly one of the central issues in evolutionary biology (Dobzhansky, 1970). A main theme of this review/perspective is to demonstrate the feasibility of studying this phenomenon at the molecular level. Reproductive isolation is generally divided into post- and premating isolation. While the evolution of premating isolation can at least be explained by the reinforcement mechanism (*i.e.*, selection against matings that yield inferior progeny; Sawyer and Hartl, 1981; Sved, 1981), postmating reproductive isolation *per se* does not appear to have any adaptive value. In fact, the two traits of postmating isolation, sterility and inviability in hybrids, are expected to be subject to strong negative selection. The pervasiveness of hybrid inviability and sterility thus presents conceptual difficulties that cannot be dispelled without invoking unsubstantiated genetic assumptions.

The pattern of postmating isolation obeys Haldane's rule remarkably well: *When in the F1 offspring of two different animal races one sex is absent, rare, or sterile, that sex is the heterogametic sex* (Haldane, 1922; see Coyne and Orr, 1989 for a review). In species with heterogametic males, Haldane's rule applies predominantly to sterility. As shown in Table I, in *Drosophila*, the numbers of documented crosses <u>for</u> and <u>against</u> this rule are 199 : 3, respectively, for sterility and 14 : 9 for inviability (compiled from Bock, 1984; Wu and Davis, 1993). In mammals,

TABLE I

Observations on F_1 of Interspecific Crosses

One sex viable		Both sexes viable, one sex fertile	
Female	Male	Female	Male
Drosophila[a]			
14	9	199	3
Mammals[b]			
0	1	25	0

Number of documented interspecific crosses in which only one sex in F_1 is viable or, if both sexes are viable, only one sex is fertile. Reciprocal crosses are counted separately.
[a] From Bock (1984); see Wu and Davis (1993).
[b] From Gray (1954).

the numbers are 25 : 0 for sterility and 0 : 1 for inviability (Gray, 1954). The patterns in both groups are consistent with the interpretation that hybrid sterility (but not inviability) in heterogametic males has evolved at a very high rate (Wu, 1992a; Wu and Davis, 1993). In this review, We will present evidence that, between three sibling species of Drosophila, genes that sterilize hybrid males upon introgression outnumber those that lead to female sterility or male inviability by many fold, perhaps as much as 10 : 1. Any attempt to explain Haldane's rule thus has to take into consideration the differences between male and female sterility as well as the differences between male sterility and inviability.

These observations suggest that spermatogenic development in Drosophila and mammals seems much more sensitive to evolutionary changes than oogenesis or somatic development. On the surface, this suggestion appears to contradict what has been learned about Drosophila somatic sex determination. It is now known that male determination represents the "default" state when all the major genes, e.g., Sxl, tra-1, and tra-2, are "off" (Hodgkin, 1990). One might expect such a state to be less sensitive to evolutionary changes than the more complex regulation leading to female development, in contrast with the actual observations. Since germ cell sex determination is very different from the somatic pathway (Schupbach, 1985; Pauli and Mahowald, 1990), the discrepancy is in part due to our inadequate understanding of the genetics of germ cell sex determination and gametogenesis.

What are the unique properties of spermatogenesis in Drosophila and mammals that make it susceptible to evolutionary changes? One of them was pointed out by Lifschytz and Lindsley (1972) in connection with the observations on X-autosome translocations, denoted T(X;A)'s by convention. Nearly all major T(X;A)'s (more than 85 of them) cause dominant male sterility in

Drosophila. These T(X;A)'s do not have any effect on female fertility and no other translocations, between autosomes, between Y and autosomes or between X and Y, cause male sterility. The effects of T(X;A)'s on spermatogenesis also appear true in mouse. These observations do not imply the prevalence of T(X; A)'s in causing hybrid sterility. (There is virtually no empirical support for that simple connection.) Rather, the relevance of these observations is in highlighting certain unique spermatogenic properties that make T(X;A)'s male sterile; for example, the "precocious X-inactivation" hypothesis Lifschytz and Lindsley proposed. Another unique property of *Drosophila* spermatogenesis is the absence of postmeiotic transcriptional regulation as the spermatids undergo one of the most dramatic transformations in cell morphology (Olivieri and Olivieri, 1965; Lindsley and Tokuyasu, 1980; Schaefer *et al.*, 1990).

The evolutionary considerations also provide a powerful intellectual incentive for studying spermatogenesis. Molecular genetic analyses of oogenesis have outpaced those on spermatogenesis, perhaps because the former is crucial for understanding embryogenic development in general, whereas spermatogenesis appears to be an end-process in itself. It is thus intriguing that spermatogenesis may have played a crucial role in the process of species formation, responsible for erecting the first reproductive barrier between incipient species. Spermatogenesis is an interesting developmental system in another evolutionary context: that of meiotic drive, or ultraselfish genes (see Werren *et al.*, 1988; Wu and Hammer, 1991 for recent reviews). This pertains to the question of the stability of Mendelian segregation and all three well-studied meiotic drive systems, two in *Drosophila* and one in mouse, are characterized by spermatogenic defects. Recently, there have been proposals to associate hybrid male sterility with meiotic drive (Hurst and Pomiankowski, 1991; Frank, 1991). Our data offer no support for such hypotheses (Johnson and Wu, 1992). There may indeed be connections between these two important evolutionary phenomena, meiotic drive and hybrid sterility; but the connections will not be obvious until we have a better understanding of the development and genetics of spermatogenesis.

In studying spermatogenesis, male sterile "mutations" can be obtained by mutagenesis, as is normally done, or, alternatively, by mixing genetic materials of two different species. Studies of the latter approach have been carried out in the past (see Coyne and Orr, 1989 for a review). Ideally, we expect these studies to answer the three questions in order. i) Is hybrid male sterility caused by discrete genetic elements and, if it is, where are they physically located? ii) What are the effects of such discrete elements on spermatogenesis and other fitness aspects?, and iii) What is the nature of their interactions with other

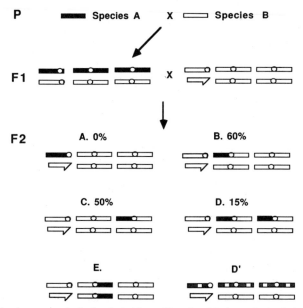

Fig. 1. F2 backcross analysis for hybrid sterility. In this hypothetical example, percent fertility of each genotype is given. This has been the most widely used approach. Although the genotypes are given as if the whole chromosomal arms had been introgressed, each type is, in fact, a collection of heterogeneous genotypes, as indicated in type D′ where the shaded areas indicate mixed origin due to the absence of visible markers. Type E, which has a homozygous autosomal introgression, cannot be recovered in such an analysis.

genes? However, our current understanding does not give a definitive answer even to the first question. It has been argued recently that complete male sterility results from the cumulative action (or complex interactions) of a large number of loci, each contributing only a small degree of male sterility (*e.g.*, Naveira and Fontdevila, 1991). Although this interpretation can be tested experimentally, the bulk of hybridization data so far are not inconsistent with it. This is because genetic analyses of hybrid sterility were usually done on F2 backcross hybrids (see Fig. 1) where the genomes are mixed with 1/4 and 3/4 from each species. At a minimum, a hybrid has 10–20% of its genes from another species. Such analyses have provided a broad base for understanding the genetics of hybrid sterility, but do not provide specific answers to the questions posed above.

An alternative approach is to introgress a small piece of chromosome from another species which carries genetic markers (Wu and Beckenbach, 1983;

Coyne and Charlesworth, 1986; Naveira and Fontdevila, 1986). When the introgression is identified with physical markers, sterility, inviability, and other more subtle aspects of fitnesses can be analyzed and attributed to the introgression itself. The introgression can be manipulated by recombination and monitored precisely by examining DNA markers. This review/perspective presents the introgression approach to the genetic analysis of postmating reproductive isolation.

MAPPING STRATEGIES

1. Basic Mapping Scheme

The scheme we have been using is an extension of the introgression protocol of Wu and Beckenbach (1983), as summarized in Fig. 2. In stage I, a

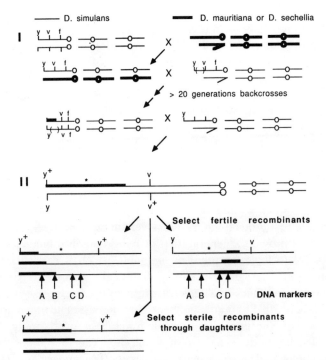

Fig. 2. The basic introgression scheme. The y v f-marked chromosome is X, drawn proportionally larger in stage II. *denotes the sterility factor and ( ) denotes a chromosome inversion with breakpoints at 2B and 8B. The same procedure is repeated for the v- and f-region.

small X chromosome segment containing the sterility factor of interest is introgressed from *Drosophila mauritiana* (or *D. sechellia*) into the genome of *D. simulans*. Repeated backcrossing to *D. simulans* males for more than 20 generations ensures a clean *D. simulans* genetic background. In stage II, males carrying a recombinant X chromosome are selected and tested for fertility. The proportion of fertile males among the recombinants permits crude genetic mapping of the factor. Each fertile recombinant X chromosome is propagated by mating males to females carrying the attached X chromosomes (male progeny in such a cross inherit the paternal X chromosome). Such recombinants are then analyzed for their sizes of introgression by using a series of DNA markers. Each DNA marker reveals a difference between *D. simulans* and *D. mauritiana* and hence the extent of introgression in each recombinant. In the example of Fig. 2, one can easily deduce the sterility factor to be between DNA markers B and C.

Most previous genetic studies of hybrid sterility (*e.g.*, Dobzhansky, 1936; see Coyne and Orr, 1989 for a review) analyzed complex genic interactions in F2 backcross hybrids (see Fig. 1). The introgression of a small chromosomal segment causing male sterility provides materials for more precise genetic, cytological, and biochemical assays. The use of an inversion [In(1) 2B-8B] also illustrates an important point: the sterility factor of interest should be kept while other factors are purged by backcrosses, yielding a clean product at the end of stage I. In(1) reduces crossovers between the *yellow* locus and the inversion to less than 0.1%. (This is measured by the presence of recombinants between *yellow* and *white* in the progeny of $In(1)y^+w^+/y\ w$ females. We also examined the polytene chromosomes of these recombinants.) Thus, the introgressed segments in the *yellow* region are not eroded by recombination during backcrosses. In other regions where inversions are not available, flanking markers are used to protect against crossovers in stage I.

The crosses provide a genetic map of hybrid sterility factors and, more importantly, create lines for physical mapping, which is carried out at two levels of precision. The coarse mapping defines the genes by recombination analyses and assigns them to a region corresponding to the size of one letter subdivision (about 200 kb) on the polytene chromosome. The fine mapping will further demarcate the genes by a small number of overlapping clones to facilitate molecular analysis. Coarse mapping, now covering about 2/3 of the X chromosome for male sterility genes from both *D. mauritiana* and *D. sechellia*, is the subject of this review.

2. DNA Marker-assisted Mapping

There are three considerations in using the nucleotide sequence itself as a marker.

1) The level of sequence divergence (and thus the resolution)

The divergence between randomly chosen genes from each of the species used (see below) is 1-2% at the nucleotide level (Coyne and Kreitman, 1986; Caccone *et al.*, 1988), which makes DNA diagnostics quite feasible. For example, every phage clone of about 15 kb is expected to detect restriction fragment length differences for most 4-bp restriction enzymes. In our studies, each clone detects interspecific differences for at least one of three restriction enzymes used. The theoretical limit of mapping the crossover points of the recombinants is 100 bp, given the 1-2% nucleotide difference. In practice, the resolution depends on the physical distance between the sterility factor and the nearest visible marker. If we assume 5cM (~ 600 kb/cM) separating the marker and the factor and 60 recombinants in that interval, then the crossover points are spaced at a 50 kb ($= 5 \times 600/60$) interval, on average. The value of 50 kb is the practical limit of mapping in this example. We now have recombinants whose crossover points are calculated to be less than 40 kb away from a sterility factor.

2) The sources and physical maps of DNA clones

In most non-*Drosophila* organisms, the construction of a physical map by linkage analysis is the labor-intensive prelude for tasks such as the Quantitative Trait Loci mapping (Helentjaris, 1987) or genetic diagnosis. In *Drosophila*, the linkage map can easily be constructed by *in situ* hybridization to polytene chromosomes. It is estimated that 91% of the *D. melanogaster* polytenized genome has been cloned (Merriam *et al.*, 1991). For example, about 50% of the X chromosome is now covered by the cosmid library of Kafatos *et al.* (1991) and more than 95% of the letter subdivisions (see legends of Fig. 4) are represented by at least one cosmid. Such a concentration of clones makes it feasible to carry out very detailed mapping of hybrid sterility genes. A very extensive coverage of the *Drosophila* genome by a yeast artificial chromosome library has also been achieved by Garza *et al.* (1989; see Merriam *et al.*, 1991 for update). These resources will also be useful (although to a less extent) for studies of *Drosophila* species outside of the *D. melanogaster* group.

3) Diagnostics of nucleotide differences

Given an introgression such as any of the recombinants of Fig. 2, we ask if it has the *D. mauritiana* or *D. simulans* gene at locus A, B, C, and so on. All the results in this review were obtained by examining the restriction fragment length differences. The procedure is reliable, but requires the isolation of quantities of genomic DNA. For rapid screening, it is thus more efficient to carry out the polymerase chain reaction (PCR) to obtain locus-specific DNA. There are at least five different methods for the detection of single nucleotide differences in the PCR products, including the Single Strand Conformation method (Orita *et al.*, 1989), the denaturing gradient gel method, the oligonucleotide hybridization method, and the oligonucleotide ligation assay (OLA, see Nickerson *et al.*, 1990). The OLA method is promising for large-scale screening because of its adaptability to automation. Its potential in clinical application should benefit the mapping project discussed in this review.

3. Choice of Species

We use *D. simulans* as the recipient species (thin line of Fig. 2) and *D. mauritiana* or *D. sechellia* (thick line) as the donor species. These species fulfill all criteria for the scheme of Fig. 2, including pervasive male sterility but little female sterility, completely homosequential chromosomes (important for generating recombinants), the availability of visible markers, the existence of X-linked inversions and attached X's and, because of their close relationships with *D. melanogaster*, a dense set of well-mapped clones. In the long run, we hope to benefit from the germ-cell transformation technology adapted for *D. simulans* (Laurie *et al.*, 1990).

The use of two donor species, *D. mauritiana* and *D. sechellia*, provides a basis for comparative studies with only modestly more effort than the use of one. There could be three patterns: 1. the introgression of a factor from *D. mauritiana* into *D. simulans* causes sterility, but the introgression of the homologous *D. sechellia* gene does not; 2. the introgression of either causes male sterility; and 3. the introgression of *D. sechellia* factor, but not its *D. mauritiana* homolog, causes sterility. Since the three species are nearly equally related (perhaps with *D. mauritiana* being slightly more ancient; see Coyne and Kreitman, 1986; Satta and Takahata, 1990), one may interpret the three patterns to mean an evolutionary change in the gene along the lineage to *D. mauritiana* (Pattern 1), *D. simulans* (2), or *D. sechellia* (3). In other words, the comparative studies enable us to infer not only a change but also the direction of the change (*i.e.*, the ancestral *vs.* the derived state).

HYBRID MALE STERILITY

1. The Forked-region Mapping

The basic scheme of Fig. 2 is, in fact, a special case for the *yellow* region

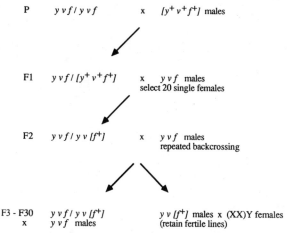

Fig. 3. The alternative introgression scheme without using flanking markers. Symbols in [ ] represent *D. mauritiana* or *D. sechellia* materials. (XX)Y females carry the attached-X chromosomes; the sons of such females inherit the paternal X.

because the *yellow* locus is at the tip of the chromosome and sterility factors can exist only on one side of the marker. If such factors exist on both sides of a marker, recombination in stage II will not yield any fertile males, creating a false impression that a factor is tightly linked. For any other markers, it is necessary to ensure that the introgression contains a sterility factor only on one side.

1) Introgression from D. mauritiana (Perez et al., 1993)

Our first set of experiments relied on the scheme of Fig. 3 which simultaneously maps the sterility factors on both sides of the *f*-marker. We started with 20 independent lines in F2 and terminated a line when some males in it became fertile starting in generation F4. Another sterile line was then split into two to keep the total number around 20. After 30 generations of backcrosses, we examined the extent of introgression in 40 fertile introgression lines and 9 sterile lines, as shown in Fig. 4. These clones were kindly provided to us by A. Chovnick (sd), E. Underwood (IM75), W. Zerges (ru), V. Corces (forked), D. Branton (KBA), M. Tanouye (Sh), L. Kalfayan (fu), and E. Stephenson (A57). Genomic DNA was digested with *Rsa* I, *Hae* III, or *Hha* I and probed with these clones.

It is clear from the pattern of Fig. 4 that a sterility factor probably exists

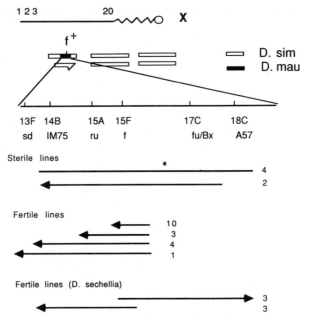

Fig. 4. RFLP-assisted mapping of a sterility factor near the marker, *f*. Lines indicate the extent of chromosomal segments introgressed from *D. mauritiana* or *D. sechellia*. Arrowheads indicate that the introgressions are beyond that point. *denotes the putative sterility factor. The number of independently derived recombinants exhibiting a comparable extent of introgression is given on the right of each line. On the top is a schematic representation of the *Drosophila* X chromosome, divided by convention into 20 regions, starting from the telomere to the base of euchromatin (next to the centromeric heterochromatin). Each region is further divided into 6 letter subdivisions (A–F). A letter subdivision is about 150–300 kb in size.

between 15F and 17C (indicated by * in Fig. 4; see the legends for the designation of chromosome regions) and, equally importantly, none exists between 15F and 13F. We therefore selected a sterile line whose introgressed segment does not extend to 13F and, hence, does not carry a sterility factor on the left of the *f*-marker. This line is subject to the recombination analysis, as shown in Fig. 5.

i) *Genetic analysis:* In total, 317 *f*⁺ *Bx* and 189 *f* *Bx*⁺ recombinant males were examined with the following results:

f^+ *Bx* males   42% fertile   ⟶   45 lines established
                            58% sterile

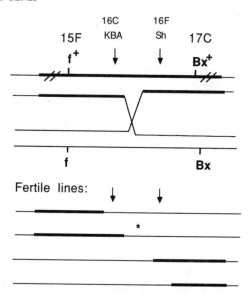

Fig. 5. Recombination analysis of the f-region factor. Thick lines represent introgressions from *D. mauritiana*. Shown in this figure are fertile genotypes.

f Bx^+ males	14% fertile	$\longrightarrow$	12 lines established
	86% sterile		

The observations confirm the existence of a sterility factor between f and Bx. The genetic data, however, cannot be reliably used to map the sterility factor because the proportions of fertile males (42% and 14%) do not add up to 100% as expected. The main reason is the presence of another sterility (Perez *et al.*, 1993). A second reason is the partial sterility of the fertile lines: when independently derived f Bx^+ or f^+Bx males are mated to attached X females, the proportion of their sons that are sterile can sometimes be as high as 30% (among 25 sons examined for each line), even though these sons are genotypically identical. Many of the sterile males must, in fact, have a "fertile genotype". The partial fertility may be caused by the presence of the *D. mauritiana* introgression on the right of Bx. Our recent observation on the recombinant f Bx sons of $[f^+]Bx/f$ Bx^+ mothers shows a much higher proportion of fertile males and is consistent with this interpretation (Perez, unpublished results). [   ] denotes the introgressed materials from *D. mauritiana*.

ii) Molecular analysis: Molecular mapping shows that many of the fertile

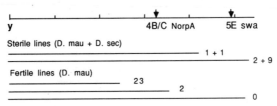

Fig. 6. RFLP-assisted mapping of a sterility factor near the marker, *y*. The marker is at the tip of the X chromosome (1A). The extent of introgression was determined by the two molecular clones. The number of lines for each type is indicated on the right.

f^+ *Bx* lines have an introgression beyond the 16C marker, but none passes 16F, suggesting the factor to be between 16C and 16F. This is supported by a fertile *f Bx*⁺ line that carries *D. mauritiana* 16F and the proximal segment. Comparing the number of crossover points between 15F–16C and 16C–16F, we map the factor to 16E. (This cytological location has recently been further confirmed by the use of another clone at 16D/E.) Since there are 16 crossover points between 16C and 16E, assuming 600 kb in that interval, we calculate the average distance between these crossover points to be less than 40 kb (600 kb/16).

2)   Introgression from D. sechellia

The last six introgressions of Fig. 4 are from *D. sechellia*. Apparently, the introgression of the 16E region from this species into *D. simulans* does not cause male sterility, suggesting that the 16E factor of *D. mauritiana* must be a derived allele from the ancestral *D. simulans/D. sechellia* type.

3)   The reciprocal introgression from D. simulans into D. mauritiana (Palopoli and Wu, unpublished results)

We also introgressed the *f-Bx* region from *D. simulans* to *D. mauritiana*, relying on the dominant marker *Bx* in selecting females for backcrosses. After 15 generations, males from 2 *f Bx* lines are fertile. Since double recombination between the markers (3.5 cM apart in our own data) is not likely, the observation shows that reciprocal introgressions are asymmetric in their sterility effect, as suggested by Wu and Beckenbach (1983).

2.   The Yellow-region (1A) (Cabot *et al.*, unpublished results)

The following results were obtained with a scheme like Fig. 3 but for the *yellow* region. Two molecular clones were used: NorpA (4B/C) and swa (5E). From Fig. 6, it is clear that a sterility factor from either *D. mauritiana* or *D. sechellia* exists between the two molecular markers. This factor from these two species may be homologous because their spermatogenic defects are also very similar (Johnson *et al.*, 1992). If one assumes that there is one crossover per

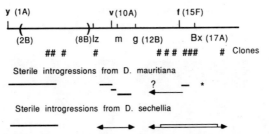

Fig. 7. Summary of X-linked introgressions causing male sterility. The positions of visible markers, inversion breakpoints and molecular clones used are all indicated on the X chromosome (the euchromatic portion). Each line denotes the smallest introgression required to yield male sterility. Arrowheads indicate that the breakpoints of the introgressions are further away. Open bar is where the sterility factor is absent. ? denotes some ambiguity about the existence of a factor.

chromosome, the factor can be mapped to 4D/E. However, since the scheme of Fig. 3 can accumulate crossovers through generations, the factor may be more proximal, between 4F and 5E.

A limited number of recombinants between y and v have also been isolated and tested by the scheme of Fig. 2. Among $[y^+]v^+$ males, 20 are fertile and 11 are sterile while, among y v males, 21 are fertile and 12 are sterile. If the factor is between 4C and 5E, roughly 1/4–1/3 and 2/3–3/4 of fertile males are expected for each class, respectively. Since the numbers are still small, it is premature at this moment to conclude a deficiency in the recovery of sterile $[y^+]v^+$ males.

3. The Vermilion-region and Summary

We have also mapped the factors in the *vermillion* region (Davis and Wu, unpublished results). Because of the relatively high density of visible markers (see Fig. 7), we could carry out the coarse mapping without relying on molecular clones. The results, together with those reported in *1* and *2*, are summarized in Fig. 7.

4. Analysis of Spermatogenic Defects
1) Cytological observations (Perez and Davis, unpublished results)

We have examined the spermatogenic defects with the phase-contrast microscopy following the procedure of Kemphues *et al.* (1982). A stage of particular interest is the so-called "onion stage" at the completion of meiosis. Spermatids of this stage are characterized by two sharply contrasted bodies lying side by side, the nucleus and the mitochondria derivative (Fig. 8a). Their presence is indicative of the completion of meiosis.

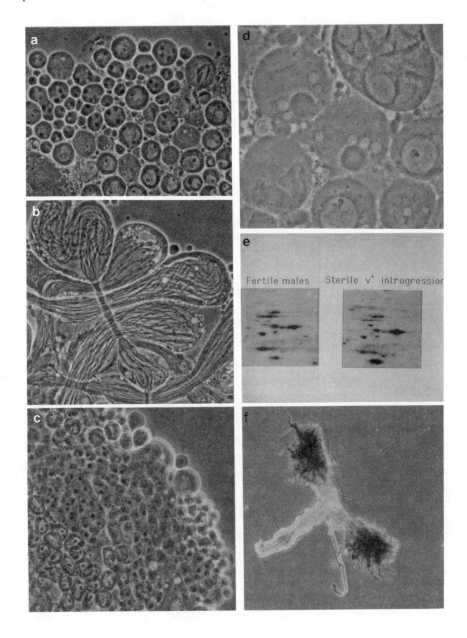

i) F1 males: All F1 males from the reciprocal crosses between *D. simulans* and *D. mauritiana* or between *D. simulans* and *D. sechellia* have onion cells (although many of the nuclei and mitochondria derivatives have an abnormal appearance). Sterility is primarily associated with spermiogenic, or postmeiotic, failure at the cytological level (Fig. 8b). This seems the most common phenotype of male sterile mutations (Lindsley and Tokuyasu, 1980). Interestingly, males with any of the introgressed sterility factors usually have a more severe phenotype despite their pure genetic backgrounds (see below).

ii) The f-Bx factor from D. mauritiana: There are two distinguishable spermatogenic defects: one with a normal onion stage like that of Fig. 8a and the other as shown in Fig. 8c. In the onion stage of the latter, no nuclei are visible. This phenotype was observed in sterile males of certain lines that have an introgression extending from *f* to beyond *Bx*. Males of other lines including those with an introgression not extending to *Bx* have a normal onion stage. The 16E factor between *f* and *Bx* appears to sterilize males without disrupting meiosis but a second factor proximal to *Bx* may aggravate the defect. We also know that some genes to the right of *Bx* cause male sterility (Perez *et al.*, 1993).

iii) The y-region factor from D. mauritiana and D. sechellia: The defects associated with this factor are very severe (and similar between the factors of the two species) as shown in Fig. 3 and Table I of Johnson *et al.* (1992). The testes are very small and mature primary spermatocytes cannot be seen; thus meiosis is not likely to happen.

iv) The v-region (distal) factor from D. mauritiana: Shown in Fig. 8d is the phenotype of the sterility factor distal to *v*, and probably proximal to *lz*. This distinctive phenotype with a fused mitochondria derivative and four separate nuclei resembles that of a *D. melanogaster* male sterile mutation, *fwd* (Fuller, 1992). The fused mitochondria derivative may have resulted from the failure to complete cytokinesis (E. Raff and M. Fuller, personal communication).

2) *Biochemical assay*

Two dimensional protein electrophoresis has been carried out. Twenty pairs of testes from males of each species, F1 hybrid males and males with an introgressed sterility factor(s) were dissected and incubated in ^{35}S-labelled methionine for 2 hr before homogenization and loading for electrophoresis (Coulthart and Singh, 1988a,b). If the profiles in sterile males are drastically

---

←Fig. 8. The spermatogenic and oogenic defects. (a) Normal onion cells. (b) Defective sperm bundles in the testes of F1 males. (c) Abnormal onion cells from sterile [*f*+] males. (d) Abnormal onion cells from sterile [*v*+] males. (e) A comparison of testis 2D protein profiles between a fertile and sterile line. (f) An example of severe oogenic defects in females of Fig. 10.

different from those of the fertile ones (due, for example, to an extensive breakdown in protein synthesis), 2D analysis will not be useful. Although this has been reported (Coulthart and Singh, 1988a,b), our results show that the profiles are usually very similar between fertile and sterile males with only a few specific changes. An example is given in Fig. 8e. In our 2D analysis, fertile males used are from the $y\ v\ f$ stock of *D. simulans* whereas the introgression males have a mixed background from several *D. simulans* strains. It is possible to make the background identical by comparing sterile recombinant males with their fertile brothers as those shown in Fig. 2.

5. Pleiotropic Effects

We ask if the introgressed male-sterility factors have pleiotropic effects in soma and in female germ cells (Johnson and Wu, 1993). The viability effects for sterile and fertile introgressions were both measured against the $y\ v\ f$ genotype as an internal standard.

1) Soma (viability)

Below is the relative viability of males (or females) carrying an introgressed male-sterile factor, relative to males (or females) with an introgression of the same region, which does not carry the factor.

	Male viability	Female viability
f-region	0.911 ($n=3,774$)	0.941 ($n=4,171$)
v-region	0.981 ($n=3,453$)	1.024 ($n=3,724$)

The results suggest that neither sterility factor has a strong negative effect on somatic functions.

2) Female fecundity

The relative fecundity of females with and without a male sterility factor (these females are from the scheme of Fig. 3):

$$y\ v\ [f^+]^*/y\ v\ f\ \text{vs.}\ y\ v\ [f^+]/y\ v\ f : 0.64\ (n=378)$$
$$y\ [v^+]^*\ f/y\ v\ f\ \text{vs.}\ y\ [v^+]f/y\ v\ f : 0.38\ (n=373)$$

where * denotes the male-sterile introgression and n is the number of females studied. The effects on female fecundity are substantial although we cannot conclude if the reduction is caused by the male sterility factor or other tightly linked genes.

6. X-Y or X-autosome Interactions?

There is a major difference between sterility caused by introgressions and sterility caused by mutations; namely, introgressed factors cause sterility only in a foreign genome, but are normal in their own background. With what do the introgressed factor interact? Various mechanisms have been proposed, including X-Y interactions (Coyne, 1985; Orr, 1987), X-autosome interactions (Dobzhansky, 1936; Muller, 1940) and Y-autosome interactions (Pantazidis *et al.*, 1992). Our approach of testing these interactions can be summarized as in Fig. 9 and Table II which address X-Y interactions. We co-introgress the Y chromosome from *D. sechellia* to suppress the sterility caused by an introgressed X chromosomal segment (type A *vs.* type B males of Fig. 9). The results for

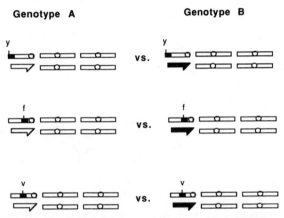

Fig. 9.   A direct test of X-Y interactions as the cause of hybrid male sterility. Solid bar represents materials from *D. sechellia* and blank bar from *D. simulans*. The difference between Genotypes A and B is the source of the Y chromosome. The description of their spermatogenic phenotypes is given in Table II.

TABLE II
Fertility and Spermatogenic Development in Males with Genotypes of Fig. 9

	Genotype A	Genotype B
y-region	Sterile	Sterile
	Small deformed testes (3.40)	Small deformed testes (3.93)
	Much reduced number of spermatocytes	Much reduced number of spermatocytes
f-region	Sterile	Sterile
	Normally-shaped testes (3.93)	Normally-shaped testes (4.90)
	Smaller spermatocytes, number reduced	Smaller spermatocytes, number reduced
v-region	Sterile	Sterile

The numbers are the mean length of testes (unit is 0.1 mm) for each genotype. The differences between genotype A and B are not significant for both *y*- and *f*-regions.

three different regions of introgression fail to confirm any rescue of fertility or spermatogenic development (Table II), suggesting that the causes of male sterility are not the interactions between any of these X-linked sterility factors and the heterospecific Y chromosome. Our observations were, in fact, more consistent with the X-autosome interaction model proposed by Dobzhansky (1936) and Muller (1940). A detailed report is given by Johnson *et al.* (1992).

7. *Meiotic Drive and Hybrid Sterility*

In our test of the X-Y interaction model, heterospecific X-Y combinations in different autosomal backgrounds were constructed. The segregation of X and Y in some of these males is expected to produce biased sex ratios under the generic model of meiotic drive causing hybrid sterility; examples of such models have recently been proposed (Frank, 1991; Hurst and Pomiankowski, 1991). We were not able to support such proposals (Johnson and Wu, 1992).

FEMALE STERILITY AND MALE INVIABILITY

In these studies of interspecific differences, we found that many traits are, in fact, affected. From the backcrosses that were carried out previously, there was no indication of female sterility or inviability (of either sex). However, we found that genic incompatibility leading to either trait is observable, if the right genotypic constructs can be obtained. The genetic basis of inviability or female sterility appears to be different from the discrete major-gene effects for male sterility.

1. *Female Sterility* (Davis and Wu, unpublished results)

Hybrid female sterility is interesting because it is so much rarer than

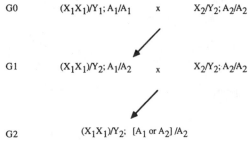

Fig. 10. Crosses that yield sterile females. The letter, A, denotes autosomes. Species 1 is *D. simulans* and (XX) is the attached X's. Species 2 is either *D. mauritiana* or *D. sechellia*. Any autosomal locus has a 50% chance of being X_1X_1; A_2A_2 in G2 females.

hybrid male sterility. This rarity is puzzling because the female pathway is thought to be more highly regulated and hence more "perturbable" as discussed in the introduction. An indication that there may be interactions affecting female fecundity can be seen in G2 of Fig. 10. Although female sterility interactions have never been reported between these three species, the reason is that, in backcrosses, females are never of such a genotype: X_1X_1; A_2A_2 (where X_1 and A_1 represent an X-linked and an autosomal gene from species 1, and X_2 and A_2 are from species 2). In other words, no X-linked loci are homozygous for genes of one species while some autosomal genes are homozygous for genes of another species. We therefore carried out the scheme of Fig. 10.

When species 2 is *D. mauritiana*, 25% (129/511) of G2 females are sterile with a spectrum of ovarian defects ranging from *ovo*-like (early germ cell death; Oliver *et al.*, 1990; see Fig. 8f) to partial atrophy. As a control, only 2% of $(X_1X_1)/Y_2$; $[A_1$ or $A_2]/A_1$ females are sterile. The observation of 25% female sterility and the variable oogenic defects suggest that there is not a single gene, or chromosomal segment, which can engender complete female sterility in a X_1X_1; A_2A_2 combination. When species 2 is *D. sechellia*, 59% (250/423) of G2 females are sterile, in comparison with the 16% sterility in $(X_1X_1)/Y_2$; $[A_1$ or $A_2]/A_1$ females.

2. *Inviability* (Cabot, Davis, and Wu, unpublished results)

Referring to the stage II of Fig. 2, we do not expect to observe any inviability associated with the recombinant chromosomes because the two parental types are viable. However, in 6 (out of about 60) independently derived $[y^+]v^+/In(1)$ y v lines, $[y^+]v^+$ males are never recovered and are apparently inviable, whereas both the non-recombinant types, $[y^+]v$ and y v^+, are viable. The genetic basis of this inviability is unknown but a detailed mapping of the recombination crossover points should be informative. One

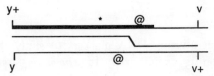

Fig. 11. An explanation for the creation of inviable recombinants. @ denotes an essential gene that is in a different position in each species. Recombinants with a crossover point between the transposed genes will yield inviable males. This model predicts the clustering of crossover points among the inviable recombinants.

possibility is given in Fig. 11 which predicts a clustering of these crossover points. An alternative explanation is the accumulation of spontaneous lethals during backcrosses.

CONCLUDING REMARKS

A broader theme in the study of hybrid sterility is the nature of interspecific differences. Male sterility, or divergence in spermatogenic development, is an obvious and important example of species difference. Many interspecific differences may not be detectable as intraspecific variation simply because the sojourn time of mutations under positive directional selection is so much shorter than neutral variations. There is indeed some evidence that intraspecific variations at nonsynonymous sites do not have the full representation of variations as might have been expected from the extent of interspecific differences (McDonald and Kreitman, 1991). The need to study interspecific differences directly, as opposed to the analysis of within-species variation, is obvious.

Genetic analysis of species differences can be done at the morphological, cytological, or molecular level. We choose to study genes underlying hybrid sterility because they can be, and should be, analyzed at all levels. Spermatogenic defects are by themselves interesting developmental questions. From an evolutionary perspective, hybrid sterility is a central issue not only because it manifests the incipient stage of speciation but also because the divergence in the spermatogenic development between incipient species is likely to have been shaped by natural selection. Although the genetic and molecular analysis of hybrid sterility has not been as detailed and refined as those on other interesting fitness-related traits such as self-incompatibility in plants (Clark, this volume) and histocompatibility in animals (Hughes, this volume), the results presented strongly suggest the feasibility of molecular analysis. This review thus represents a first step toward studying speciation at the molecular level.

SUMMARY

A new approach utilizing DNA markers to study the genetic basis of reproductive isolation is summarized. In *Drosophila*, hybrid male sterility appears to have evolved at a very high rate and is usually the first indication of species divergence. Genetics of hybrid sterility or inviability has previously been analyzed only at the level of whole chromosome or large chromosomal region. In this new approach, small pieces of chromosomes are introgressed from one species into another by repeated backcrosses. The extent of introgression can be known by examining the RFLP patterns of the hybrids at a series of DNA markers whose physical locations are known. On the X chromosome alone, we found six genes that can sterilize hybrid males between a pair of very closely related *Drosophila* species. Such male sterility genes have nearly complete penetrance, many of them having a distinct spermatogenic phenotype. Some of these hybrid sterility genes have now been defined with sufficient precision to facilitate molecular cloning. On the other hand, genetic divergence between species appears to have a much smaller and less discrete effect on female sterility. The approach of this review represents a first step toward studying the molecular nature of speciation.

Acknowledgment

The work was supported by NIH grants (GM39902 and HG 00005), an NDSS fellowship to C.-I Wu and a Sloan Foundation postdoctoral fellowship to E. Cabot.

REFERENCES

Bock, I.R., 1984. *Evol. Biol.* **18**: 41-70.
Caccone, A., G.D. Amato, and J.R. Powell, 1988. *Genetics* **118**: 671-683.
Coulthart, M.B. and R.S. Singh, 1988a. *Mol. Evol. Biol.* **5**: 167-181.
Coulthart, M.B. and R.S. Singh, 1988b. *Mol. Evol. Biol.* **5**: 182-191.
Coyne, J.A., 1985. *Nature* **314**: 736-738.
Coyne, J.A. and B. Charlesworth, 1986. *Heredity* **57**: 243-246.
Coyne, J.A. and M. Kreitman, 1986. *Evolution* **40**: 673-691.
Coyne, J.A. and H.A. Orr, 1989. In *Speciation and Its Consequences*, edited by D. Otte and J. Endler, pp. 180-207. Sinauer Press, Sunderland, Massachusetts.
Dobzhansky, Th., 1936. *Genetics* **21**: 113-135.
Dobzhansky, Th., 1970. *Genetics of the Evolutionary Process.* Columbia Univ. Press, New York.
Frank, S.A., 1991. *Evolution* **45**: 262-267.
Fuller, M.T., 1992. Spermatogenesis. (in preparation).
Garza, D., J.W. Ajioka, D.T. Burke, and D.L. Hartl, 1989. *Science* **246**: 641-646.
Gray, A.P., 1954. *Mammalian Hybrids.* Commonwealth Agricultural Bureau. Farnham Royal, England.
Haldane, J.B.S., 1922. *J. Genet.* **12**: 101-109.
Helentjaris, T., 1987. *Trends Genet.* **3**: 217-221.
Hodgkin, J., 1990. *Nature* **344**: 721-728.
Hurst, L.D. and A. Pomiankowski, 1991. *Genetics* **128**: 841-858.
Johnson, N.A., D.E. Perez, E.L. Cabot, H. Hollocher, and C.-I Wu, 1992. *Nature* (in press).
Johnson, N.A. and C.-I Wu, 1992a. *Genetics* **130**: 507-511.
Johnson, N.A. and C.-I Wu, 1993. *Am. Natur.* (submitted).

Kafatos, F.C., C. Louis, C. Savakis, D.M. Glover, M. Ashburner, A.J. Link, I. Siden-Kiamos, and R.D.C. Saunders, 1991. *Trends Genet.* **7**: 155-161.

Kemphues, K.J., T.C. Kaufman, R.A. Raff, and E.C. Raff, 1982. *Cell* **31**: 655-670.

Laurie, C.C., E.M. Heath, J.W. Jacobson, and M.S. Thomson, 1990. *Proc. Natl. Acad. Sci. U.S.A.* **87**: 9674-9678.

Lifschytz, E. and D.L. Lindsley, 1972. *Proc. Natl. Acad. Sci. U.S.A.* **69**: 182-186.

Lindsley, D.L. and K.T. Tokuyasu, 1980. In *The Genetics and Biology of Drosophila*, Vol. 2, edited by M. Ashburner and T.R.F. Wright, pp. 226-294. Academic Press, New York.

McDonald, J.H. and M. Kreitman, 1991. *Nature* **351**: 652-654.

Merriam, J.,M. Ashburner, D.L. Hartl, and F.C. Kafatos, 1991. *Science* **254**: 221-225.

Muller, H.J., 1940. In *The New Systematics*, edited by J.S. Huxley, pp. 185-268. Clarendon Press, Oxford.

Naveira, H. and A. Fontdevila, 1991. *Heredity* **67**: 57-72.

Nickerson, D.A., R. Kaiser, S. Lappin, J. Stewart, L. Hood, and U. Landegren, 1990. *Proc. Natl. Acad. Sci. U.S.A.* **87**: 8923-8927.

Oliver, B., D. Pauli, and A.P. Mahowald, 1990. *Genetics* **125**: 535-550.

Olivieri, G. and A. Olivieri, 1965. *Mutat. Res.* **2**: 366-380.

Orita, M., Y. Suzuki, T. Sekiya, and K. Hayashi, 1989. *Genomics* **5**: 874-879.

Orr, H.A., 1987. *Genetics* **116**: 555-563.

Pantazidis, A.C., V.K. Galanopoulos, and E. Zouros, 1992. *Genetics* (submitted).

Pauli, D. and A.P. Mahowald, 1990. Germ-line sex determination in *Drosophila melanogaster*. *Trends Genet.* **6**: 259-264.

Perez, D.E., C.-I Wu, N.A. Johnson, and M.-L. Wu, *Genetics* (submitted).

Satta, Y. amd N. Takahata, 1990. *Proc. Natl. Acad. Sci. U.S.A.* **87**: 9558-9562.

Sawyer, S. and D.L. Hartl, 1981. *Theor. Popul. Biol.* **19**: 261-273.

Schaefer, M., R. Kuhn, F. Bosse, and U. Schaefer, 1990. *EMBO J.* **9**: 4519-4525.

Schupbach, T., 1985. *Genetics* **109**: 529-548.

Sved, J.A., 1981. *Genetics* **97**: 197-215.

Werren, J.H., U. Nur, and C.-I Wu, 1988. *Trends Ecol. Evol.* **3**: 297-302.

Wu, C.-I, 1992a. *Evolution* (in press).

Wu, C.-I and A.W. Davis, *Am. Natur.* (submitted).

Wu, C.-I and A.T. Beckenbach, 1983. *Genetics* **105(1)**: 71-86.

Wu, C.-I and M.F. Hammer, 1991. In *Evolution at the Molecular Level*, edited by R.K. Selander, A.G. Clark, and T.S. Whittam, pp. 177-203. Sinauer Press, Sunderland, Massachusetts.

Mitochondrial Transmission in *Drosophila*

ETSUKO T. MATSUURA

Department of Biology, Ochanomizu University, Bunkyo-ku, Tokyo 112, Japan

The transmission of mitochondrial genomes is unique in that multiple copies of the genome are present within a cell. The transmission of mitochondrial DNA (mtDNA) as a population can be treated by population genetics theory (Birky, 1978). Extensive theoretical studies have been carried out on organelle transmission to provide some clarification of the evolutionary processes of mitochondrial genomes (*e.g.*, Chapman *et al.*, 1982; Clark, 1984; Takahata, 1984; Birky *et al.*, 1989). Experimental systems for verification of these models, however, have been limited to several organisms.

Most transmission studies have been made using unicellular systems such as somatic cultured cells and yeast (for reviews, see Birky, 1983, 1991; Gingold, 1988). In multicellular systems, transmission analysis is possible in heteroplasmic individuals present in natural populations and their progeny, using mtDNA variation as genetic markers of mitochondria. The transmission of mtDNA has been investigated in *Drosophila mauritiana* (Solignac *et al.*, 1984), *Gryllus firmus* (Rand and Harrison, 1986), and Holstein cows (Hauswirth and Laipis, 1982; Laipis *et al.*, 1988). The heteroplasmic state appears to be retained stably for many generations, particularly in insects, and the effect of selection on mtDNA transmission, if it exists, is considered to be very small.

In *Drosophila*, heteroplasmy can be constructed by transplanting germ plasm. By this technique, foreign mtDNA can be introduced into germline cells even from a different species (Matsuura *et al.*, 1989; Tsujimoto *et al.*, 1991), and the heteroplasmic state is retained in later generations. We previously reported selective transmission of mtDNA at the heteroplasmic state induced by intra- and interspecific transplantation of germ plasm (Matsuura *et al.*, 1990, 1991; Tsujimoto *et al.*, 1991). In an extreme case, foreign mtDNA was found to completely replace endogenous mtDNA (Niki *et al.*, 1989; Tsujimoto *et al.*, 1991). The selection depended on temperature (Matsuura *et al.*, 1991).

Clarification of the mechanisms for selective mtDNA transmission is essential to an understanding of the evolution of the mitochondrial genome and its relationship to the nuclear genome. The present *Drosophila* system should facilitate the study of mtDNA transmission in multicellular systems and provide additional insight into these studies. In this paper, mtDNA transmission in artificially heteroplasmic *D. melanogaster* is examined. The mechanisms for selective mtDNA transmission and its significance in mtDNA evolution are discussed.

ESTABLISHMENT OF MITOCHONDRIAL DNA HETEROPLASMY BY GERM-PLASM TRANSPLANTATION

Transplantation of germ plasm was carried out according to Matsuura *et al.* (1989). With several strains of *D. melanogaster* as recipients, *D. melanogaster*, *D. simulans*, *D. mauritiana*, and *D. sechellia* were used as germ-plasm donors. Two

TABLE I

Strain Combinations in Germ-plasm Transplantation Using *D. melanogaster* as a Recipient

Recipient	Donor	Source/reference
$bw;e^{11}$	*D. melanogaster* (HJ6)	a
$bw;e^{11}$	*D. simulans* (SL61)	b
$bw;e^{11}$	*D. simulans* (SI259)	c
$bw;e^{11}$	*D. mauritiana* (g20)	a
$bw;st$	*D. melanogaster* (L149)	d
$bw;st$	*D. simulans* (SL61)	d
$bw;st$	*D. mauritiana* (g20)	d
$bw;st$	*D. sechellia* (SS78)	e
L149	*D. melanogaster* (*bw;st*)	d
y;bw;st	*D. mauritiana* (g20)	d

a) Matsuura *et al.* (1989).
b) Niki and Matsuura (unpublished data).
c) Yamamoto, Niki, and Matsuura (unpublished data).
d) Tsujimoto *et al.* (1991).
e) Tsujimoto and Matsuura (unpublished data).

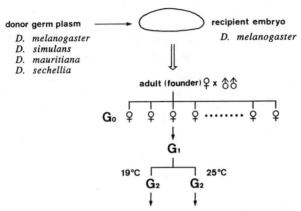

Fig. 1. Procedure for the transplantation of germ plasm and establishment of heteroplasmic lines. Isofemale lines were established from individual inseminated G_0 females and their mitochondrial types determined based on progeny (G_1). The progeny of each heteroplasmic line (G_2 or G_3) was divided into two and each was maintained at a different temperature. (From Matsuura, 1991)

types of coexisting mtDNA, derived from a recipient and donor, can be distinguished by their restriction patterns. Table I presents the strain combinations for transplantation. For each combination, four to six heteroplasmic lines were established and maintained at two different temperatures by dividing the progeny into two at the second or third generation (Fig. 1). Thus, essentially the same line was examined for mtDNA transmission at 19°C and 25°C.

The proportion of donor-derived mtDNA within a line was monitored in later generations, using mtDNA extracted from flies which were used as parents for the next generation, as previously described (Matsuura *et al.*, 1989).

SELECTIVE TRANSMISSION OF MITOCHONDRIAL DNA IN HETEROPLASMIC LINES

Germ plasm was successfully transplanted into recipient eggs and heteroplasmic individuals were obtained from both intra- and interspecific combinations, when using *D. melanogaster* as a recipient species (Matsuura *et al.*, 1989; Tsujimoto *et al.*, 1991; Yamamoto, Niki, and Matsuura, unpublished data). Donor-derived mtDNA was retained with endogenous mtDNA under the novel nuclear genome in later generations. However, one of the two types of mtDNA was selectively transmitted in all cases, irrespective of intra- and interspecific combinations.

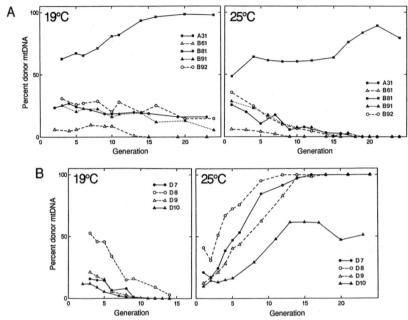

Fig. 2.   Changes in the proportion of donor mtDNA in heteroplasmic lines using the *bw;e^11* strain as a recipient. (A) Heteroplasmy for the intraspecific combination, using the HJ6 strain of *D. melanogaster* as a donor. (B) Heteroplasmy for the interspecific combination, using the g20 strain of *D. mauritiana* as a donor. (From Matsuura *et al.*, 1991)

The rates of change in the proportion of donor-derived mtDNA were shown to depend on temperature (Fig. 2). In intraspecific combinations of *D. melanogaster*, the proportion of donor mtDNA decreased more rapidly at 25°C in one case (Fig. 2A) and at 19°C in another, depending on the strain used (Matsuura *et al.*, 1991; Tsujimoto *et al.*, 1991). In interspecific combinations, the proportion of donor mtDNA decreased when using as donors *D. simulans* (Tsujimoto *et al.*, 1991) and *D. sechellia* (Tsujimoto and Matsuura, unpublished data) and increased when using *D. mauritiana* as a donor (Tsujimoto *et al.*, 1991). In an extreme case out of 10 combinations, donor mtDNA derived from *D. mauritiana* completely replaced endogenous *bw;e^11* mtDNA by the 15th generation at 25°C, but was lost by the 14th generation at 19°C (Fig. 2B; Matsuura *et al.*, 1991).

The intensity of selection in mtDNA transmission was estimated by a simple model based on genic selection, assuming sufficiently large numbers of mtDNA molecules and mitochondria within a cell and random sorting of mitochondria to daughter cells. When the differential replication of donor and recipient mtDNA for every fly generation is in the proportion of $1+s$ to 1, respectively, the increase in the proportion of donor mtDNA ($\Delta p/\Delta t$) becomes $sp(1-p)$. Applying the overlapping generation model of genic selection, an equation for the proportion of donor mtDNA at time t (p_t) is given in Crow and Kimura (1970) as

$$p_t = \frac{1}{1 + \dfrac{(1-p_0)}{p_0} e^{-st}},$$

where p_0 is the initial proportion of donor mtDNA in a germline cell. Estimates

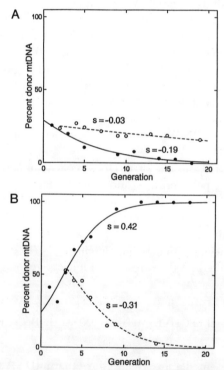

Fig. 3.   Examples of theoretical curves for changes in the proportion of donor mtDNA expected from the model. Observed proportions for the A31 and D8 lines given in Fig. 2 are re-plotted for 19°C (○) and 25°C (●) in graphs (A) and (B), respectively. The progeny of the lines at 25°C were divided into two at the second generation for A31 and at the third generation for D8, and each was maintained at 19°C and 25°C. Estimates of the selection coefficient (s) for both temperatures are also shown.

of s were made by fitting the equation to the data by the non-linear least squares method. Changes in the proportion of donor mtDNA were found to fit the expectations well. Figure 3 shows examples of the theoretical curve obtained from the equation for the intra- and interspecific combinations at 19°C and 25°C. The intensity of selection estimated as a selection coefficient clearly depended on temperature (Matsuura, Niki, and Chigusa, submitted).

The involvement of mtDNA replication in selective mtDNA transmission is suggested from the effects of the A+T-rich region on the intensity of selection. Using as donors *D. simulans* and *D. mauritiana* which possess the same type of mtDNA (Solignac *et al.*, 1986), heteroplasmy was constructed using *D. melanogaster* as a recipient (Table I). The nucleotide sequences of the coding region for the two kinds of mtDNA were shown to be almost identical (Satta and Takahata, 1990), but the A+T-rich regions presumably differ from each other considerably in their sequences since the A+T-rich region containing the replication origin is most variable in the genome within and between species (Fauron and Wolstenholme, 1980a, b). Changes in the proportion of donor-derived mtDNA were examined and the intensity of selection was shown to differ for the two kinds of donor mtDNA (Yamamoto *et al.*, unpublished data). This indicates that some sequences within the A+T-rich region are responsible for the selection in mtDNA transmission. The details of the results will be reported elsewhere.

Another factor responsible for mtDNA replication is the nuclear genome which encodes DNA polymerase γ and other proteins necessary for mtDNA replication. The influence of the nuclear genome on the intensity of selection was also examined in two kinds of heteroplasmy by reciprocal combinations of transplantation (Table I), that is, using the same two types of mtDNA under the two different nuclear genomes. The intensity of selection against the same type of mtDNA differed for the two kinds of nuclear genomes (Tsujimoto *et al.*, 1991).

IMPLICATIONS OF SELECTIVE MITOCHONDRIAL DNA TRANSMISSION IN EVOLUTION

The results presented above clearly show that mtDNA transmission at the heteroplasmic state in *Drosophila* is selective and controlled by both mitochondrial and nuclear genomes. Interactions between some sequences within the A+T-rich region and nuclearly encoded proteins may be temperature-dependent at the initiation of mtDNA replication. The involvement of mtDNA replication in selective mtDNA transmission has also been suggested in unicellular organisms such as yeast and mammalian somatic cultured cells (*e.g.*, Shay and Ishii, 1990; Zweifel and Fangman, 1991). The temperature-dependence of selection in the present *Drosophila* system should contribute to the clarification of the molecular mechanisms for selection in mtDNA transmission.

It is important to understand the evolution of mtDNA in relation to the nuclear genome. Functional coordination between mitochondrial and nuclear genomes is apparently essential to the maintenance and transmission of mitochondria as well as to the respiratory function. The mechanisms for regulating mitochondrial genomes by the nuclear genome, including replication and expression of mtDNA, should have been evolved along with the evolution of mtDNA. The possible transfer of many genes from ancient mitochondrial genomes to the nuclear genome should have also been involved in the process of the establishment of regulating mechanisms. Recently, a nuclear gene which controls the biased transmission of mtDNA in yeast was identified and suggested to be involved in mtDNA replication (Zweifel and Fangman, 1991). Evolutionary studies on these kinds of genes in various organisms should be conducted.

Selection in mtDNA transmission may be one significant evolutionary force on the mitochondrial genome. A model of within-generation selection has been proposed and shows a strong effect on the substitution rate (Takahata, 1984). In higher animals and insects, mtDNA variation can be introduced through paternal leakage, in addition to mutation. The incomplete maternal transmission of mtDNA has recently been reported in backcross hybrids in *Drosophila* (Kondo *et al.*, 1990, 1992) and mouse (Gyllensten *et al.*, 1991). The effects of selection, in addition to those of random drift, should thus be regarded as more important to the evolution of mtDNA than previously considered.

Further studies, both experimental and theoretical, on the selective mtDNA transmission in *Drosophila* are necessary to understand organelle transmission mechanisms. Clarification of the molecular mechanisms for the regulation of the selection will reveal heretofore unknown aspects of the evolution of mtDNA in eucaryotes.

SUMMARY

The transmission of mtDNA was investigated in heteroplasmic lines of *Drosophila* established by germ-plasm transplantation. Using several strains of *D. melanogaster* as recipients and *D. melanogaster* and its sibling species as germ-plasm donors, heteroplasmy was constructed in various donor-recipient combinations. Heteroplasmic lines for each combination were examined for changes in the proportion of donor-derived mtDNA at different temperatures for more than 10 generations. Selective transmission of one type of mtDNA was observed in all combinations. Changes in the proportion of donor mtDNA fitted the expectations from a simple model based on genic selection, assuming differential replication of mitochondria and/or mtDNA. The intensity of selection was found to depend on the temperature at which the heteroplasmic lines were maintained. The A+T-rich region in mtDNA and nuclear genome were both apparently involved as factors determining temperature-dependent selection in mtDNA transmission. The significance of selective transmission in mtDNA evolution is discussed.

Acknowledgments

The author is grateful to Prof. S.I. Chigusa, Ochanomizu University, for his encouragement and valuable comments during the course of this study. She thanks Dr. Y. Niki, Ibaraki University, for collaboration in the transplantation experiments and discussions and Ms. Y. Tsujimoto and N. Yamamoto for providing unpublished data. Thanks are also due to Drs. N. Takahata, National Institute of Genetics, and K. Matsuura, Tokyo Metropolitan University, for their continuous interest and comments. This work was supported in part by Grants-in-Aid for Scientific Research from the Ministry of Education, Science and Culture of Japan.

REFERENCES

Birky, C.W., Jr., 1978. *Annu. Rev. Genet.* **12**: 471–512.
Birky, C.W., Jr., 1983. *Science* **222**: 468–475.
Birky, C.W., Jr., 1991. In *Evolution at the Molecular Level*, edited by R.K. Selander, A.G. Clark, and T.S. Whittam, pp. 112–134. Sinauer Associates, Sunderland, Massachusetts.
Birky, C.W., Jr., P. Fuerst, and T. Maruyama, 1989. *Genetics* **121**: 613–627.
Chapman, R.W., J.C. Stephens, R.A. Lansman, and J.C. Avise, 1982. *Genet. Res.* **40**: 41–57.
Clark, A.G., 1984. *Genetics* **107**: 679–701.
Crow, J.F. and M. Kimura, 1970. *An Introduction to Population Genetics Theory*. Harper & Row, New York.
Fauron, C.M.-R. and D.R. Wolstenholme, 1980a. *Nucleic Acids Res.* **8**: 2439–2452.
Fauron, C.M.-R. and D.R. Wolstenholme, 1980b. *Nucleic Acids Res.* **8**: 5391–5410.
Gingold, E.B., 1988. In *Division and Segregation of Organelles*, edited by S.A. Boffey and D. Lloyd, pp. 149–170. Cambridge University Press, Cambridge.
Gyllensten, U., D. Wharton, A. Josefsson, and A.C. Wilson, 1991. *Nature* **352**: 255–257.
Hauswirth, W.W. and P.J. Laipis, 1982. *Proc. Natl. Acad. Sci. U.S.A.* **79**: 4686–4690.
Kondo, R., E.T. Matsuura, and S.I. Chigusa, 1992. *Genet. Res.* **59**: 81–84.

Kondo, R., Y. Satta, E.T. Matsuura, H. Ishiwa, N. Takahata, and S.I. Chigusa, 1990. *Genetics* **126**: 657-663.

Laipis, P.J., M.J. Van de Walle, and W.W. Hauswirth, 1988. *Proc. Natl. Acad. Sci. U.S.A.* **85**: 8107-8110.

Matsuura, E.T., 1991. *Jpn. J. Genet.* **66**: 683-700.

Matsuura, E.T., S.I. Chigusa, and Y. Niki, 1989. *Genetics* **122**: 663-667.

Matsuura, E.T., S.I. Chigusa, and Y. Niki, 1990. *Jpn. J. Genet.* **65**: 87-93.

Matsuura, E.T., Y. Niki, and S.I. Chigusa, 1991. *Jpn. J. Genet.* **66**: 197-207.

Niki, Y., S.I. Chigusa, and E.T. Matsuura, 1989. *Nature* **341**: 551-552.

Rand, D.M. and R.G. Harrison, 1986. *Genetics* **114**: 955-970.

Satta, Y. and N. Takahata, 1990. *Proc. Natl. Acad. Sci. U.S.A.* **87**: 9558-9562.

Shay, J.W. and S. Ishii, 1990. *Anticancer Res.* **10**: 279-284.

Solignac, M., J. Génermont, M. Monnerot, and J.-C. Mounolou, 1984. *Mol. Gen. Genet.* **197**: 183-188.

Solignac, M., M. Monnerot, and J.-C. Mounolou, 1986. *J. Mol. Evol.* **23**: 31-40.

Takahata, N., 1984. *Genet. Res.* **44**: 109-116.

Tsujimoto, Y., Y. Niki, and E.T. Matsuura, 1991. *Jpn. J. Genet.* **66**: 609-616.

Zweifel, S.G. and W.L. Fangman, 1991. *Genetics* **128**: 241-249.

Genetic Polymorphisms and Recombination in Natural Populations of *Escherichia coli*

THOMAS S. WHITTAM AND STACEY E. AKE

Institute of Molecular Evolutionary Genetics, Department of Biology, Pennsylvania State University, University Park, PA 16802, U.S.A.

In contrast to sexually reproducing organisms in which the meiotic processes of crossing-over and independent assortment produce recombinant genotypes each generation, bacteria reproduce asexually and create recombinant genotypes only through mechanisms of gene transfer, such as conjugation, transformation, and transduction, which are independent of cell division. The fact that such mechanisms are prevalent among different taxonomic groups of bacteria and the documentation of numerous cases of transfer of genes between distantly related bacteria have engendered the notion that strains of bacteria exchange genetic information in a promiscuous fashion.

This, however, cannot be true, because there are barriers to the wholesale exchange of genes between bacterial lines. For example, mismatch repair systems inhibit recombination between highly divergent homologous sequences and can disrupt exchange between strains, as in the case of *Escherichia coli* and *Salmonella typhimurium* (Rayssiguier *et al.*, 1989). Restriction enzyme systems also prevent the acquisition of foreign DNA by protecting against bacteriophage infections, which may result in a reduction of gene flow within and between bacterial populations (Daniel *et al.*, 1988).

With the opposition of mechanisms that promote and those that hinder the transfer of genetic information between strains, it is unclear to what extent bacterial populations represent gene pools, analogous to species of sexually reproducing organisms, or clone mixtures in which the genotype of each clone is essentially a closed system isolated from other similar ones. These two extremes of population structure strongly contrast in their evolutionary properties. In a common gene pool, genotypic allele combinations are ephemeral in evolutionary time because they are constantly created and broken down by recombination. Individual alleles are the units of evolutionary change, with natural selection and genetic drift causing shifts in gene frequency and gene substitution. In a clone mixture, however, genotypes (cell lines) can persist for many generations and are capable of changing only through mutation and rare recombination events. If recombination occurs at a very low rate, the occasional appearance of an advantageous mutation results in the replacement of cell lines, a phenomenon referred to as periodic selection (Atwood *et al.*, 1951). Because the clonal genotype is the unit of evolutionary change, the random sampling of lines resulting from stochastic extinction or periodic selection profoundly reduces the effective population size and the amount of genetic variation in a population (Kubitschek, 1974; Maruyama and Kimura, 1980; Levin, 1981).

Our objective in this chapter is to address the following general question. In asexually reproducing bacteria that have mechanisms for the horizontal transfer of genes between strains, what role does recombination play in the generation of genotypic diversity in natural populations? To answer this question, we will examine patterns of enzyme polymorphism in wild strains of *E. coli* and estimate the magnitude of the evolutionary parameters of mutation and recombination obtained for a natural population. We will also consider recent studies of DNA polymorphism in genes encoding structural proteins to assess amounts of variation and rates of recombination at the nucleotide level.

For present purposes, we will distinguish three classes of recombination events. The distinctions are based not on the mechanism of gene transfer (conjugation, transduction, or transformation) or the details of the molecular pathways for recombining DNA molecules (see Weinstock, 1987), but rather, on the outcome of recombination events from a population genetic perspective. Events in each class involve the transfer of genetic material between strains and result in the production of recombinant genotypes. *Assortative recombination* events are those in which new chromosomal genotypes are produced by the reshuffling of existing alleles into new combinations. *Intragenic recombination* involves substitutions of pieces of genes, shorter in length than a cistron, that generate novel mosaic alleles and, thus, new genotypes. Finally, *additive recombination* occurs when genetic elements integrate to form a composite genotype that is the sum of the recombining molecules. One important type of additive event is the acquisition of genes from other bacterial species.

LEVELS OF ASSORTATIVE RECOMBINATION

To assess the rates at which extant alleles in natural populations of bacteria are reshuffled by gene transfer to create recombinant genotypes, a method of monitoring allelic variation at multiple loci in the chromosomal genome is required. Multilocus enzyme electrophoresis (Selander *et al.*, 1986) has been widely applied to bacterial populations because it readily permits an assessment of allelic variation at multiple loci in the large numbers of isolates required for population analysis.

In application to *E. coli* populations, the study of protein polymorphisms has provided evidence for a clonal structure. The clonal hypothesis originated from early observations of identical phenotypes, in such variable traits as serotype and biotype, among *E. coli* strains recovered from separate outbreaks of disease (Ørskov *et al.*, 1976; Ørskov and Ørskov, 1983). Selander and Levin (1980) extended the clone concept to the *E. coli* species as a whole, based on the repeated recovery of isolates with identical multilocus enzyme genotypes. This finding was incompatible with high rates of assortative recombination, given the large number of alleles observed per locus. Further evidence for a clonal population structure in *E. coli* was the demonstration of extensive linkage disequilibrium for many enzyme loci (Selander and Whittam, 1983; Whittam *et al.*, 1983a, b). The frequencies of multilocus genotypes in natural populations departed significantly from those expected under a model of random association (Whittam *et al.*, 1983a), and the modal genotype, the genotypic combination of the most common alleles, was not observed in a sample of more than 1,600 isolates (Ochman *et al.*, 1983).

Given a clonal population structure in which assortative recombination is infrequent, the analysis of all sets of genetic characteristics that are representative of the genome should yield similar genetic relationships among isolates. Numerous studies have demonstrated a concordance between phylogenetic relationships inferred from protein polymorphisms and other characteristics (Ochman and Selander, 1984; Miller and Hartl, 1986; Whittam and Wilson, 1988; Ørskov *et al.*, 1990; Arbeit *et al.*, 1990). But, at the same time, discordant phylogenies, based on comparisons of different genes or different character sets, have been used as evidence of past recombination (Selander *et al.*, 1987; DuBose *et al.*, 1988; Biserčič *et al.*, 1991; Dykhuizen and Green, 1991). Evidence relevant to the clonal population structure model has been summarized at various stages in reviews by Hartl and Dykhuizen (1984), Achtman and Pluschke (1986), Selander *et al.* (1987), and Young (1989).

1.   Genetic Variation in the E. coli Population of Children in 3 Mexican Villages

To gain insight into the genetic structure of *E. coli* populations in regions where enteric diseases are a major public health problem, we have undertaken a study of bacteria recovered from the inhabitants of three remote, rural villages in Campeche, Mexico, on the eastern coast of the Yucatan peninsula. These are people of Mayan descent who are self-sufficient practitioners of subsistence agriculture, animal husbandry, and wild game hunting. Water supplies are seasonally variable and generally inadequate, and solid human waste is disposed of in fields or in walled-in yards adjacent to the houses. At the time of our study, village families averaged 5–6 children per household, of whom half under the age of 4 were experiencing diarrheal episodes as well as other symptoms of infection and intestinal parasitism.

Thus far in our study, we have examined electrophoretic variation in 20 enzymes in 317 *E. coli* isolates cultured from stool samples of 13 children in 3 villages. Equating electromorphs of an enzyme with alleles at the corresponding structural locus, we have estimated the amount of genetic variability in the bacterial population within a village in terms of genetic diversity, or "virtual heterozygosity" (Kimura, 1983), which is defined for a single locus as one minus the sum of the squared allelic frequencies. The average genetic diversity across multiple loci provides an estimate of the probability, for the average locus, that two genes sampled at random from a population are different alleles. For the flora of hosts within villages, the average genetic diversity per locus ranges from 0.31 to 0.45, with a mean for the pooled sample of 0.41 (Table I). These values are slightly lower than those that have been reported for *E. coli* enteric populations from humans in other localities, which range from 0.44 in Finland to 0.54 in Massachusetts (Selander *et al.*, 1987).

TABLE I

Genetic Variation among *E. coli* Isolates from Children in 3 Mexican Villages

Sample	No. infants	No. isolates	Average no. alleles	Average genetic diversity ± SE
Village 1	4	92	3.5	0.348 ± 0.059
Village 2	4	104	3.2	0.309 ± 0.061
Village 3	5	121	4.3	0.445 ± 0.053
Total	13	317	5.2	0.414 ± 0.053

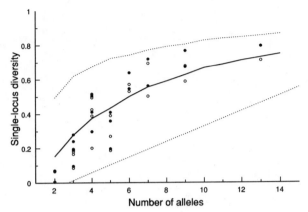

Fig. 1. Fit of observed single-locus genetic diversity to the values predicted from the sampling theory of neutral alleles (Ewens, 1979). Solid circles mark estimates of total diversity, open circles mark average within-village diversity; the solid line is drawn through the points predicted for a sample size of 317; and the dotted lines demarcate the 95% confidence region.

We compared the levels of single-locus genetic diversity with those predicted under the hypothesis of the strict neutrality of molecular polymorphisms (Ewens, 1972). According to the infinite alleles model of neutral mutation, an evolving population will achieve a steady state in which the production of new alleles by mutation is balanced by the random loss of pre-existing alleles. At steady state, the expected virtual heterozygosity, $E(h)$, is given by $M/(1+M)$, where, for a haploid organism, $M = 2N_e\nu$ and N_e and ν are the effective population size and neutral mutation rate, respectively (Kimura, 1983). Figure 1 shows the relationship between the number of alleles and the genetic diversity observed for each locus, calculated both in the pooled sample and as the average within villages. The observed single-locus diversities closely fit the values predicted from the sampling theory of neutral mutations in an equilibrium population (Ewens, 1972, 1979; Watterson, 1978). Because all values fall within the range of neutrality, there is no reason to reject the hypothesis that allelic variation detected by protein electrophoresis within the *E. coli* population is effectively neutral.

2. Single-locus Diversity and the Number of Codons per Gene

The equilibrium level of virtual heterozygosity in a haploid population is, as mentioned above, a function of the neutral mutation parameter M. This parameter can be roughly interpreted as twice the long-term average number

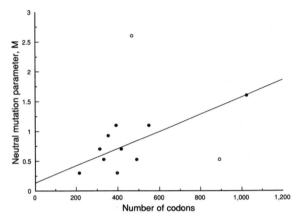

Fig. 2.   Relationship between the magnitude of the neutral mutation parameter and protein size (number of codons) for 12 enzyme encoding genes.

of neutral variants introduced into a population each generation, and its level is related to factors that influence the effective population size (N_e) and the neutral mutation rate per locus (ν). One prediction of this model is that for a population at evolutionary equilibrium there should be a positive correlation across loci between the neutral mutation rate and the magnitude of M and, hence, the single-locus diversity at equilibrium.

To test this prediction, we have examined the relationship between estimates of M, obtained from the single-locus diversities for the Mexican village data and the number of codons specifying an enzyme, which were determined from sequences in GenBank. An assumption underlying this analysis is that the rate of neutral mutation at the nucleotide or codon level is roughly constant so that a large protein will have a higher mutation rate than a smaller protein. For 12 enzyme-encoding loci in which both genetic diversities and nucleotide sequences were available, there is a positive relationship between the number of codons and the estimate of M (Fig. 2). Although the linear regression is not significant for all 12 enzymes, it is significant ($R^2 = 0.615$, $F = 12.8$, $p < 0.01$) when two outlying points are omitted from the analysis (see below). This correlation indicates that much of the interlocus variance in the level of neutral variation is accounted for by the overall size of the protein. Similar relationships between protein subunit size and heterozygosity based on electrophoretically detected polymorphisms have been reported in several groups of eukaryotic species (Nei, 1987).

There are two enzymes in *E. coli*, alcohol dehydrogenase (ADH) and

6-phosphogluconate dehydrogenase (PGD), that strongly depart from the linear relationship between M and the number of codons (see Fig. 2). In the case of ADH, the amount of neutral variation is less than that predicted from the size of the protein. One explanation for this observed deficit of variation is that conventional starch-gel electrophoresis does not resolve many of the amino acid variants of this protein. It is noteworthy that ADH has a high frequency of "null" phenotypes (*i.e.*, no detectable enzyme activity under standard electrophoretic conditions) in *E. coli* populations; for example, Caugant *et al.* (1983) reported a frequency of 0.26 for nulls at ADH in a sample of 268 isolates from Sweden. Such nulls may represent a heterogeneous collection of variants that, when pooled together into a single allelic class for data analysis, yields an underestimate of the actual amount of genetic variation at the locus. A second possibility is that the ethanol dehydrogenase activity assayed as ADH is produced by a protein other than that encoded by the chromosomal gene *adhE*, a gene that specifies a protein of 892 amino acids in length. Although there are several distinct proteins described in *E. coli* with ethanol dehydrogenase activity, we have been unable to distinguish them based on differences in substrate specificity. Thus, the reason for the departure of ADH from the size-diversity correlation remains unresolved.

In contrast to ADH, PGD is nearly 3 times more variable than expected on the basis of the number of codons in the gene. This excess variation presumably reflects an inflated rate of generation of new electrophoretic variants at the *gnd* locus, and comparative nucleotide sequencing of *gnd* from multiple strains of *E. coli* has revealed clustered polymorphisms at this locus, suggesting that intragenic recombination has played a significant role in generating allelic variation in this gene (Sawyer, 1989).

3. Estimating Levels of Assortative Recombination

In an effort to extend Ewen's sampling theory of neutral alleles to multiple loci, Hedrick and Thomson (1986) used Monte Carlo simulations to examine the sampling properties of measures of linkage disequilibrium between alleles at two loci (A and B) in a population at equilibrium. For haploid chromosome $A_i B_j$ that occurs in a population at frequency x_{ij}, the commonly used measure of linkage disequilibrium is

$$D_{ij} = x_{ij} - p_i q_j$$

where p_i and q_j are the observed frequencies of alleles A_i and B_j. To measure the total disequilibrium between all alleles at two loci, Hedrick and Thomson employed, among others, the standardized measure

$$Q^* = \frac{\Sigma_{ij}D_{ij}{}^2/p_iq_j}{(k-1)(l-1)}$$

where k and l denote the number of alleles at loci A and B, respectively. Under the null hypothesis that $D_{ij}=0$ (*i.e.*, no associations between alleles), the product of the numerator of the above equation and the sample size is distributed approximately as chi-square with $(k-1)(l-1)$ degrees of freedom. Thus, Q^* provides a measure of the intensity of association between alleles at two loci that is roughly independent of sample size (Hedrick and Thomson, 1986).

Through extensive computer simulations, Hedrick and Thomson (1986) tabulated the distribution of Q^* for different sample sizes, numbers of alleles, and levels of recombination for populations at equilibrium. Using data from 100 *E. coli* isolates recovered from humans in Sweden and characterized for variation at 12 enzyme loci (Caugant *et al.*, 1983), they found that most of the mean values of Q^* for 45 locus-pairs fell between the theoretical values of $C=$ 10 and $C=100$, where $C=2N_ec$ (c being the fraction of recombinant genotypes produced each generation). Furthermore, the distribution of D_{ij} for six pairs of loci with three alleles at one locus and six alleles at the other was close to the theoretical distribution for neutral alleles for $C=10$. This analysis clearly demonstrated that the neutral recombination parameter, C, was greater than 0 in this sample of *E. coli* from Sweden.

We have used the methods of Hedrick and Thomson (1986) to obtain estimates of C between polymorphic protein loci in isolates from children in Mexican villages. For the 190 possible pairwise comparisons between 20 polymorphic enzyme loci, 43 (23%) of the Q^* values were significantly different from zero when compared to a chi-square distribution with $df=(k-1)(l-1)$, indicating that alleles for many locus-pairs were in significant linkage disequilibrium. Overall, the average $Q^*=0.104$, with a standard error of 0.208. The large standard error reflects both the sampling variation in Q^* for locus-pairs with the same numbers of alleles at each locus and the effect of pooling values for locus-pairs with different numbers of alleles. Although the calculation of Q^* involves division by the degrees of freedom, which partially corrects for the dependence on number of alleles, the expected value of Q^* for neutral alleles in a fixed sample size decreases as the number of alleles per locus increases (see Table 2 of Hedrick and Thomson, 1986).

The variation in Q^* among locus-pairs is illustrated in Fig. 3, in which the mean estimate is plotted against df for each locus-pair. A power function fit to the means shows an inverse relationship between mean Q^* and df, with a strong rise in Q^* when $df < 20$. A similar power function fit through the points representing one standard error above and below each mean shows extensive sampling variation around each mean, especially for low values of df. To assess the significance of mean Q^*, we randomized the multilocus genotypes of isolates on the computer by sampling without replacement the alleles at each locus and constructing new genotypes. The sampling scheme removed associations between alleles at different loci but preserved the overall sample size and the allele frequencies, thus mimicking a sample from a population with free recombination ($C \gg 1$). The mean Q^* for the randomized genotypes across 190 locus-pairs was 0.0038 ± 0.0095, which is $< 5\%$ of the observed mean and indicates that observed Q^* is significantly greater than expected for the random association of alleles expected with free recombination. Also shown in Fig. 3 are the theoretical values for increments of C between $C = 0$ (no recombination) and $C = 100$ (virtually free recombination). The theoretical curves for C equal to 1 and to 10 bracket most of the observed means as well as the observed curve.

In sum, the analysis of linkage disequilibrium reveals that the neutral recombination parameter, C, is greater than 0 in the samples of *E. coli* recover-

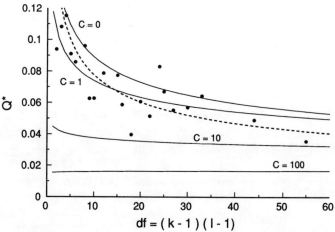

Fig. 3. Linkage disequilibrium statistic Q^* as a function of the degrees of freedom for 190 locus-pairs. The points are means for locus-pairs with the same df. The dashed line is a power function fit to the observed points, and the solid lines are power functions fit to the theoretical values generated by Hedrick and Thomson (1986) for a sample size of 100 and 4 levels of recombination. The point for $df = 1$ has a $Q^* = 0.30$ and is not shown in the figure.

ed from children in Mexico. However, the amount of recombination is less than that found by Hedrick and Thomson's (1986) analysis of a Swedish population, suggesting that the rate of assortative recombination in the average *E. coli* population is within an order of magnitude of the neutral mutation rate.

NUCLEOTIDE POLYMORPHISMS AND INTRAGENIC RECOMBINATION

1. Loci Examined

To assess the contribution of intragenic recombination to the generation of new alleles, we have examined nucleotide polymorphism data for 11 protein-encoding chromosomal genes. Allelic variation in nucleotide sequence has been determined for several loci among wild strains of *E. coli*, many of which are in the *E. coli* reference (ECOR) collection (Ochman and Selander, 1984). Milkman and Crawford (1983) reported variation in nucleotide sequences of a region of the *trp* operon in 12 strains of *E. coli* sampled from nature. We have used data for 11 strains that includes 1,113 bases from the 3′ end of *trpC* through the 5′ end of *trpB*. Dykhuizen and Green (1986) determined nucleotide sequences for 766 bases in the middle of *gnd* (which encodes PGD) for eight strains of the ECOR collection; and, more recently, Biserčić *et al.* (1991) sequenced the entire reading frame of 1,407 bases of the *gnd* locus in 10 ECOR isolates. Stoltzfus *et al.* (1988) studied nucleotide polymorphism in a region upstream of the *trp* operon that includes two open reading frames (ORF2 and ORF3) of 501 and 507 bases, respectively (Stoltzfus *et al.*, 1988). DuBose *et al.* (1988) examined variation in nucleotide sequence of the *phoA* locus, including a segment of 1,416 bases that encodes alkaline phosphatase. Finally, Nelson *et al.* (1991) determined the nucleotide sequence of 924 bases of the coding region of the *gapA* locus, which encodes glyceraldehyde-3-phosphate dehydrogenase, among 13 natural *E. coli* isolates.

In addition to these sequences, Milkman and Bridges (1990) have examined DNA polymorphism in 15 regions of the *E. coli* genome by four-base cutting restriction enzyme analysis for all 72 strains of the ECOR collection. R. Milkman has kindly provided high-resolution restriction site maps for five genes, as follows: *nirR* (fumarate and nitrate reduction regulatory protein), *fumB* (anaerobic class I fumarase), *tonB* (sensitivity to T1 phages), *topA* (DNA topoisomerase I), and *fumA* (fumarase).

2. Estimates of the Neutral-mutation Parameter M

From the available sequence data, we tabulated the number of polymorphic nucleotide sites (S) and the average number of differences between pairs of sequences (k), estimated $M(S)$ and $M(k)$, and from these calculated Tajima's D statistic (Tajima, 1989). The values of M range from 65.04 for $M(S)$ for *gnd* to 2.36 for $M(k)$ for the *gap* alleles (Table II). Under the neutral hypothesis, the two estimates of M should not be significantly different, and none of the six D values falls outside of the 95% confidence intervals listed in Tajima's Table 2. However, the negative D for the *gap* alleles slightly exceeds the lower 90% limit for $n=13$ (Tajima's Table 2), and lies in the direction suggesting that some variants may be weakly deleterious. Overall, the differences between the two estimates of M are insignificant and do not deviate from those expected under the neutral mutation hypothesis.

A similar analysis of alleles of five coding regions based on DNA polymorphisms detected by four-cutter restriction enzymes provides two estimates of M that can be used in Tajima's test. In this case, $M(S)$ was estimated by $p/\sum_i^{n-1}(1/i)$, in which n is the number of sequences and p is the proportion of

TABLE II

Estimates of the Neutral Mutation Parameter (M) and Tajima's D Statistic from Nucleotide Sequences

Locus	No. sequences	No. bases	No. polymorphic sites	$M(S)$	$M(k)$	D
gnd (middle)	7	766	100	40.82	40.57	−0.04
gnd	10	1,407	184	65.04	60.73	−0.33
ORF2	5	501	35	16.80	15.50	−0.58
phoA	8	1,416	63	24.30	21.93	−0.53
ORF3	4	507	16	8.73	8.83	0.12
trpBC	11	1,113	49	16.73	11.44	−1.49
gapA	13	924	12	3.87	2.36	−1.59

TABLE III

Estimates of the Neutral Mutation Parameter (M) and Tajima's D Statistic from Polymorphic Restriction Sites[a]

Locus	No. sequences	No. cut sites	No. polymorphic sites	$M(S)$	$M(k)$	D
nirR	48	35	18	4.06	5.01	0.75
fumB	47	55	32	7.25	6.12	−0.53
tonB	44	51	23	5.29	5.49	0.13
topA	49	51	23	5.16	4.37	−0.50
fumA	51	63	31	6.89	6.19	−0.34

[a] Restriction site data provided by R. Milkman.

polymorphic nucleotide sites (Nei, 1987). We obtained p from the number of polymorphic restriction sites using Hudson's (1982) equation (7). $M(k)$ was calculated from π by the method of Nei (1987). For each of the five genes listed in Table III, the paired estimates of M do not differ significantly, as indicated by the magnitude of Tajima's D, and thus do not provide evidence for departure from the predictions of the neutral mutation hypothesis.

3. Nucleotide Diversity

Although M is an important quantity for determining amounts of neutral variation at the level of the locus, it is also necessary to have a measure of variation at the nucleotide level. At the nucleotide level, we calculated p, the proportion of polymorphic nucleotide sites ($= S/T$, where T is the total number of nucleotides) and π, the nucleotide diversity for 11 genes. Nucleotide diversity yields the probability that two randomly sampled sequences from a population have different nucleotide bases at the average site and can be interpreted as a measure of the probability that a site is heterozygous in a randomly selected diploid individual (Nei, 1987). For haploid genomes, such as those of bacteria and eukaryotic mitochondria, nucleotide diversity can be thought of as the virtual heterozygosity per site. For the 11 *E. coli* genes, the overall level of genetic variation at the nucleotide level ranges 10 fold, from the extensive polymorphism observed for *gnd* to the low level of polymorphism observed across *gap* alleles (Table IV). The two measures of variation are highly correlated, with a linear regression coefficient of 3.1 and $R^2 = 0.91$ (Fig. 4).

TABLE IV
Variation at Nucleotide Sites

Locus	Proportion poly-morphic sites (p)	Nucleotide diversity (π)
gnd (middle)	0.131 ± 0.066	0.0454 ± 0.0259
gnd	0.131 ± 0.058	0.0388 ± 0.0208
nirR	0.087 ± 0.026	0.0321 ± 0.0157
fumB	0.103 ± 0.032	0.0292 ± 0.0146
ORF2	0.070 ± 0.042	0.0248 ± 0.0158
tonB	0.073 ± 0.022	0.0221 ± 0.0109
fumA	0.082 ± 0.024	0.0218 ± 0.0108
topA	0.073 ± 0.022	0.0181 ± 0.0090
phoA	0.045 ± 0.022	0.0136 ± 0.0077
ORF3	0.032 ± 0.022	0.0131 ± 0.0093
trpBC	0.044 ± 0.020	0.0093 ± 0.0052
gapA	0.013 ± 0.006	0.0024 ± 0.0016

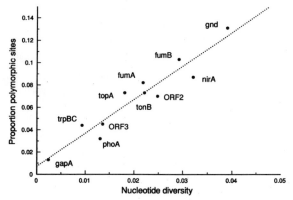

Fig. 4. Nucleotide sequence variation among alleles for 11 genes in *E. coli*.

For *gnd* we have two estimates of variation, one based on the partial sequences of Dykhuizen and Green (1986) and another on the entire coding region sequenced by Biserčić *et al.* (1991). The two data sets have identical *p* values, but the middle region shows a somewhat greater level of nucleotide diversity than that of the gene as a whole (Table IV). Overall, π ranges from 0.039 for *gnd* to 0.002 for *gapA* with an average across 11 genes of 0.035. For comparison, the average nucleotide diversity per locus in *E. coli* is 5 times greater than that observed at ADH in *Drosophila melanogaster* (Kreitman, 1983).

4. Estimates of the Neutral-recombination Parameter C

Hudson (1987) has proposed an estimator of the recombination parameter $C\ (=2N_ec)$ that is based on the variance in the number of site differences between pairs of sequences in a random sample. This estimator can be applied to both nucleotide sequence and restriction site data. Through extensive Monte Carlo simulations, Hudson (1987) found that his formulation provides an estimate near the true value of recombination for a wide range of values, although large samples are required to obtain reliable estimates.

Using the nucleotide sequences for five *E. coli* genes, we applied Hudson's method to estimate *C* for comparisons both across genes and to the neutral mutation parameter *M*. In this case, *C* refers to the amount of recombination among sites *within* a gene each generation. The values of *C* when all polymorphic nucleotide sites were considered range from 23.2 for *phoA* to 331.3 for *gapA* genes (Table V). The ratio *C/M* falls into a narrow range of 1 to 5, except for the *gapA* sequences, which have an extraordinarily high ratio (Table V). For four of the five loci, both the estimates of *C* and *C/M* remain relatively

TABLE V
Recombination Parameters Estimated from Nucleotide Sequences

Locus	All sites		Synonymous sites	
	C	C/M	C	C/M
gnd	147.3	2.4	137.3	2.7
ORF2	49.5	3.2	36.2	2.6
phoA	23.2	1.1	21.5	1.2
ORF3	45.7	5.2	54.7	8.2
gapA	331.3	140.4	16.9	9.7

TABLE VI
Recombination Parameters Estimated from Polymorphic Restriction Sites

Locus	C	C/M
nirR	4.2	0.8
fumB	5.8	1.0
tonB	11.2	2.0
fumA	18.5	3.0
topA	33.6	7.7

unchanged when only synonymous site differences are used in Hudson's
formulation. But the fact that the ratio for the *gapA* alleles drops to about 10
when only synonymous sites are considered indicates that the nonsynonymous
differences between certain alleles at this locus account for the inflated former
ratio. A similar analysis of restriction site polymorphisms for five loci is summarized in Table VI. The values of C range from about 4 to 34, with C/M
ranging from about 1 to 8.

The greatest absolute estimate of C based on synonymous sites is for the
gnd alleles sequenced by Biserčić *et al.* (1991). The magnitude of this parameter
suggests that intragenic recombination occurs at a rate at least 100 times
greater than the reciprocal of the effective population size. Indeed, this estimate
of C may be low because Hudson's estimator fails to converge when applied to
Dykhuizen and Green's (1991) *gnd* data, which indicates that the true value of
the recombination parameter is very large. Sawyer (1989) also invoked
intragenic recombination at this locus to explain extensive nonrandom clustering of polymorphic sites along the gene. In addition, Dykhuizen and Green
(1991) found evidence for intragenic recombination at the *gnd* locus in the
discordance between phylogenetic trees of strains inferred from *trp* and from
phoA sequences.

ADDITIVE RECOMBINATION

Additive recombination can result from a variety of gene transfer events, including integration of bacteriophages and plasmids, and translocation of insertion sequences and transposable elements (Schwesinger, 1977). We will not review the numerous studies of additive recombination here; but instead, we will focus attention on genes that are suspected to have recently originated from outside the *E. coli* genome, thus representing interspecific transfer events.

Evidence for Recent Acquisition of Foreign Genes

There are many examples of gene exchanges between different species of prokaryotes, some involving very distantly related organisms (recently reviewed by Mazodier and Davies, 1991). One intriguing example that points to the recent incorporation of foreign genes into the *E. coli* population is the discovery that 13% of wild *E. coli* strains produce an unusual satellite DNA that is synthesized by bacterial reverse transcriptase (Herzer *et al.*, 1990). This satellite DNA, called multicopy single-stranded DNA (msDNA), is unusual because it consists of a single-stranded DNA linked to an RNA sequence by a 2′,5′-phosphodiester linkage (Dhundale *et al.*, 1987) and exists in many copies per cell. The genes that encode reverse transcriptase and msDNA are linked together in a "retron", which, according to Inouye and Inouye (1991), represents a primitive type of retroelement.

Although uncommon in the *E. coli* population, msDNA-producing strains occur in several highly divergent clusters of strains, suggesting that retrons have been independently acquired in separate phylogenetic lineages. In fact, in one case, a retron is part of a cryptic prophage (Inouye *et al.*, 1991) and, in another case, a retron is contained within a larger unique sequence that was probably integrated into the *E. coli* genome by transposition or phage integration (Hsu *et al.*, 1990). The hypothesis that these retrons had diverged in other species of bacteria before being transferred into *E. coli* is supported by the observation that the average codon adaptation index (Sharp and Li, 1987) for the reverse transcriptase (RT) genes of two msDNA-producing strains is 0.17 (Herzer *et al.*, 1990), a value lower than those reported for any of 165 *E. coli* genes (Sharp and Li, 1987). Such a low degree of codon adaptation strongly argues that retrons are foreign to *E. coli* and were transferred relatively recently and independently into different lineages. However, the phenotypic effects of msDNA and RT expression and the adaptive benefits that are conferred to host cells, if any, remain to be elucidated.

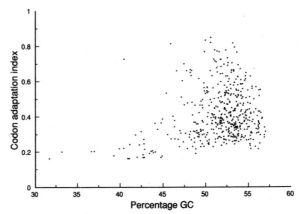

Fig. 5.   Scatter plot of 500 genes based on % GC content and the codon adaptation index.

As this example illustrates, genes that have recently been transferred into the *E. coli* genome often show atypical patterns of codon usage. To estimate the extent of recent transfers representing gene flow into the *E. coli* population, we have tabulated the percentage GC content and the codon adaptation index (CAI) for 500 coding *E. coli* sequences obtained from GenBank (Fig. 5). Most of the loci fall into a cloud of points between 47–57% GC content and range in CAI from 0.20–0.85. However, approximately 6% of the points have % GC < 45 and CAI < 0.37. These genes are atypical for the *E. coli* genome and include, for example, the coding regions of the msDNA retron, the *erm* genes encoding high-level resistance to macrolide-licosamide-streptogramin antibiotics (Brisson-Noël *et al.*, 1988), and hemolysin-encoding genes (Felmlee *et al.*, 1985), all of which are considered to have been recently transferred into *E. coli* from distantly related organisms.

EVOLUTIONARY PERSPECTIVES

1.   Levels of Neutral Variation

One of the goals of population genetics is to refine estimates of the parameters of the evolutionary process in natural populations. For the *E. coli* population comprising the normal enteric flora of children from Mexican villages, our estimates of the neutral mutation parameter fell in the range of $M = 0.5$–1.0 based on the amounts of genetic diversity detected by enzyme electrophoresis. Because $M = 2N_e\nu$, there is direct inverse relationship between the effective population size and the rate of neutral mutation, so that information about the size of one parameter can be used to infer the size of the other.

For example, if we use a working value of $M = 1$ for *E. coli* populations, then, under the assumption that the rate of electrophoretically detectable mutation is on the order 10^{-7} per locus per generation, the effective population size is on the order 10^7. This value for N_e seems unreasonably small considering that the standing crop of *E. coli* cells may be as large as 10^{20} (Milkman and Stoltzfus, 1988). Hence, we arrive at a paradox; either N_e is much smaller than seems biologically reasonable or if N_e is larger, say on the order of 10^{10} (Milkman and Stoltzfus, 1988), then the mutational production of electrophoretic variants occurs at a rate much slower than expected.

Resolution of this paradox comes from consideration of the effects of periodic selection and random extinction of cell lines under low rates of recombination. Kubitschek (1974) recognized that in asexual haploid populations accumulated neutral variation would be lost during periodic selection events. Nei (1976) suggested that the long-term effective size of *E. coli* populations may be smaller than 10^{10} because of frequent extinction of lines. Maruyama and Kimura (1980) found that when local extinction and recolonization occur frequently, the effective size of a haploid population consisting of independent cell lines can be profoundly reduced, to the order of 10^7, if local extinction occurs as frequently as every thousand generations. Levin (1981) has also demonstrated that, in principle, low rates of periodic selection and gene exchange purge neutral variation in bacterial systems. He concludes that under reasonable values of population density, turnover rate, and gene transfer, the fate of neutral alleles in populations of *E. coli* would be similar to the fate of neutral alleles in populations of sexually reproducing species with small effective sizes.

A second consideration is that the production of electrophoretically detectable alleles, which involves the rate of nonsynonymous mutation, may be lower in bacteria than in eukaryotic species. For instance, Ochman and Wilson (1987) report that the rate of nonsynonymous substitution in bacterial genes is roughly one-fifth the rate observed in mammals. To estimate the rate of electrophoretically detectable mutations, we compared the sequences of 62 homologous genes in *E. coli* and *S. typhimurium* and tabulated the number of charged amino acid differences. There are on average 9.3 charged amino differences per locus which translates to 4.6 charge differences per locus per lineage. With the assumption that protein electrophoresis detects all charge differences and using Ochman and Wilson's estimate that *E. coli* and *S. typhimurium* diverged 140 million years ago, we obtain a rate of charged amino acid substitution of 3.2×10^{-8} per year or 3.2×10^{-10} per generation (assuming 10^2 generations per year).

Thus, the equilibrium levels of protein polymorphism in natural populations of *E. coli* are consistent with a neutral mutation parameter of $M = 1$. This parameter represents the product of the rate of mutation of electrophoretic variants per locus, which we estimate could be as low as 3×10^{-10} per generation, and the effective population size, which concomitantly could be as large as 3×10^9. This estimate of the effective population size is remarkably close to that obtained by Ochman and Wilson (1987) based on the silent substitution rate. A value of $M = 1$ is also consistent with a frequency of about 5% for the repeated recovery of the same electrophoretic type (based on 20 enzyme loci) from a population (*cf.* Maynard Smith, 1991).

2.   Estimates of Recombination Parameters

The level of assortative recombination between loci, based on our application of the Hedrick and Thomson's method, falls within an order of magnitude ($1 < C < 10$) of the neutral mutation parameter for the Mexican village data. This means that the rate of assortative recombination in natural *E. coli* populations could be as low as 10^{-10}–10^{-9} per generation. Although estimates of these parameters are crude at best, with large sampling errors, they can be refined by studies of variation at the DNA level.

For the 11 genes that have been examined thus far for DNA polymorphism, the amount of variation per locus ranges within an order of magnitude, as measured by the probability that a site is polymorphic between alleles (values from 0.013 for *gapA* to 0.130 for *gnd*). From the results of Tajima's test, we cannot reject the hypothesis that the variation at these loci is neutral. The level of intragenic recombination measured by Hudson's estimator of C indicates that the rate of formation of mosaic alleles is several times greater than the neutral mutation rate, suggesting that intragenic recombination plays a significant role in generating new alleles at a locus.

Intragenic recombination appears to have been particularly important in the evolution of variation at the *gnd* locus. In comparison to the levels of protein polymorphisms for other enzymes, this locus has 2–3 times more variation than expected based on the size of the gene (Fig. 2). And although a highly variable enzyme in *E. coli* populations, PGD is not a good indicator of overall genetic relatedness of strains as reflected in the low correlation with genetic distance (Selander *et al.*, 1987).

One factor that may account for the unusually high variability of the *gnd* locus is the indirect effect of natural selection operating on the nearby *rfb* gene cluster, a region involved in the synthesis of the lipopolysaccharide O antigen. In *E. coli* antigenic expression is highly variable, and strains with particular O antigens are prevalent in certain diseases of humans and domestic animals, while occurring non-randomly in the normal enteric bacteria of these hosts (Ørskov *et al.*, 1977). Biserčić *et al.* (1991) have suggested that the close proximity of *gnd* and *rfb* inhibits genetic drift at the *gnd* locus, presumably because of the action of selection on antigenic variation. One possibility is that recombinants involved in the *gnd-rfb* region have selective advantages and increase in frequency under certain conditions that favor strains with specific O antigens. By attaining higher frequencies, these variants are less likely to be lost by random drift or through turnovers caused by periodic selection.

3. Clonal Frames

To accommodate recombination, natural selection, and mutation in the evolution of *E. coli* populations, Milkman and colleagues have developed a model in which a *clonal frame*, or specific chromosomal background, is driven to high frequency by an advantageous mutation. While this clonal frame is increasing and spreading geographically, it accumulates neutral mutations and replaces bits of its genome through gene transfer. The degree to which the clonal frame remains intact depends, of course, on the relative rates of these processes.

Although the parameters of the clonal frame model are not all directly comparable to the neutral parameters discussed here, we can compare the estimates of the intragenic recombination rates. Milkman and Bridges (1990) estimate a replacement rate of 5×10^{-12} per nucleotide per generation, which

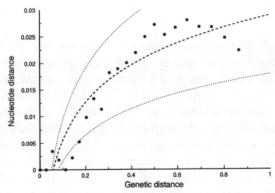

Fig. 6. Correlation of genetic divergence between strains based on distance estimated from 38 enzymes and restriction maps of five genes. Curves mark power functions fit to the mean±SE for pairs of strains at the same genetic distance.

is on the same order of magnitude as our estimate of 10^{-9}–10^{-10} per locus, given that the average locus is about one thousand nucleotides.

Under the clonal frame model, correlations in the degree of divergence based on different gene loci should be maintained for many generations, if replacements are infrequent and involve short segments of DNA. To assess the integrity of clonal frames, we compared divergence estimates between pairs of strains based on genetic distance calculated from both allelic variation at 38 enzyme loci and restriction site variation based on five other loci (Fig. 6). The average distances between the pairs of strains can be fit by an increasing power function that reflects the overall correlation of these two measures of genetic distance. Although there is substantial variation around the function, the positive correlation of distance based on the two sets of genes suggests that information about the overall extent of genetic divergence of genomes is retained over long periods of time.

As we have shown in this chapter, most of the data fit the predictions of the neutral model, and both assortative and intragenic recombination have played a significant role in generating genotypic variation. A key point is that there is a large variance in the level of genetic diversity and the rate of recombination among various genes. In particular, the *gnd* locus stands out among the genes examined here as having undergone a high rate of intragenic recombination. Given this variance across loci, estimates of C and other evolutionary parameters that pertain to the entire genome should not be based on studies of single genes, any more than phylogenies of organisms (or genomes) should be based on studies of single loci.

SUMMARY

The study of enzyme polymorphisms has shown that natural populations of *E. coli* harbor extensive genetic diversity that is organized into a limited number of genetically distinct clones. Such a population structure suggests that the rate of recombination between strains in nature is low. Here, we distinguish three types of past recombination events: assortative recombination in which already existing alleles in the population are assorted into new genotypic combinations; intragenic recombination in which pieces of genes recombine to generate new alleles and thus new genotypes; and additive recombination events in which genes are transferred from other bacterial species into the *E. coli* genome. Distributions of linkage disequilibrium coefficients between allozymes indicate that assortative recombination occurs about as frequently as mutation in the evolutionary divergence of multilocus enzyme genotypes.

Comparisons of the patterns of nucleotide polymorphisms among alleles of 10 different protein-encoding genes reveal that many polymorphic sites are clustered, as reflected in the variance of the number of site differences, which suggests that intragenic recombination and neutral mutations contribute equally to the generation of new alleles. Finally, a survey of the joint distribution of GC content and the codon adaptation index for 500 *E. coli* genes indicates that about 6% of genes show aberrant values presumably as a result of their having been recently acquired by transfer from other species of bacteria.

REFERENCES

Achtman M. and G. Pluschke, 1986. *Annu. Rev. Microbiol.* **40**: 185–210.

Arbeit, R.D., M. Arthur, R. Dunn, C. Kim, R.K. Selander, and R. Goldstein, 1990. *J. Infect. Dis.* **161**:220–235.

Atwood, K.C., L.K. Schneider, and F.J. Ryan, 1951. *Proc. Natl. Acad. Sci. U.S.A.* **37**: 146–155.

Biserčić, M., J.Y. Feutrier, and P.R. Reeves, 1991. *J. Bacteriol.* **173**: 3894–3900.

Brisson-Noël, A., M. Arthur, and P. Courvalin, 1988. *J. Bacteriol.* **170**: 1739–1745.

Caugant, D.A., B.R. Levin, G. Lidin-Janson, T.S. Whittam, C. Svanborg Eden, and R.K. Selander, 1983. *Prog. Aller.* **33**: 203–227.

Daniel, A.S., V.F. Fuller-Pace, D.M. Legge, and N.E. Murray, 1988. *J. Bacteriol.* **170**: 1775–1782.

Dhundale, A., B. Lampson, T. Furuichi, M. Inouye, and S. Inouye, 1987. *Cell* **51**: 1105–1112.

DuBose, R.F., D.E. Dykhuizen, and D.L. Hartl, 1988. *Proc. Natl. Acad. Sci. U.S.A.* **85**: 7036–7040.

Dykhuizen, D.E. and L. Green, 1986. *Genetics* **113**: s71.

Dykhuizen, D.E. and L. Green, 1991. *J. Bacteriol.* **173**: 7257–7268.

Ewens, W.J., 1972. *Theor. Popul. Biol.* **3**: 87–112.

Ewens, W.J., 1979. *Mathematical Population Genetics*. Springer-Verlag, Berlin.

Felmlee, T., S. Pellett, and R.A. Welch, 1985. *J. Bacteriol.* **163**: 94–105.

Hartl, D.L. and D.E. Dykhuizen, 1984. *Annu. Rev. Genet.* **18**: 31–68.

Hedrick, P.W. and G. Thomson, 1986. *Genetics* **112**: 135–156.

Herzer, P.J., S. Inouye, M. Inouye, and T.S. Whittam, 1990. *J. Bacteriol.* **172**: 6175–6181.

Hsu, M-Y., M. Inouye, and S. Inouye, 1990. *Proc. Natl. Acad. Sci. U.S.A.* **87**: 9454–9458.

Hudson, R.R., 1982. *Genetics* **100**: 711–719.

Hudson, R.R., 1987. *Genet. Res. Camb.* **50**: 245–250.

Inouye, M. and S. Inouye, 1991. *TIBS* **16**: 18–21.

Inouye, S., M. Sunshine, E. Six, and M. Inouye, 1991. *Science* **252**: 969–971.

Kimura, M., 1983. *The Neutral Theory of Molecular Evolution*. Cambridge University Press, Cambridge.

Kreitman, M., 1983. *Nature* **304**: 412–417.

Kubitschek, H.E., 1974. In *Evolution in the Microbial World*, edited by M.J. Carlile and J.J. Skehel, pp. 105–130. Cambridge University Press, London.

Levin, B.R., 1981. *Genetics* **99**: 1–23.

Maruyama, T. and M. Kimura, 1980. *Proc. Natl. Acad. Sci. U.S.A.* **77**: 6710–6714.

Maynard Smith, J., 1991. *Proc. R. Soc. Lond. B.* **245**: 37–41.

Mazodier, P. and J. Davies, 1991. *Annu. Rev. Genet.* **25**: 147–171.

Milkman, R. and M.M. Bridges, 1990. *Genetics* **126**: 505–517

Milkman, R. and I.P. Crawford, 1983. *Science* **221**: 378–380.

Milkman, R. and A. Stoltzfus, 1988. *Genetics* **120**: 359–366.

Miller, R.D. and D.L. Hartl, 1986. *Evolution* **40**: 1–12.

Nei, M., 1976. *Trends Biochem. Sci.* **1**: N247–N248.

Nei, M., 1987. *Molecular Evolutionary Genetics.* Columbia University Press, New York.

Nelson, K., T.S. Whittam, and R.K. Selander, 1991. *Proc. Natl. Acad. Sci. U.S.A.* **88**: 6667–6671.

Ochman, H. and R.K. Selander, 1984. *Proc. Natl. Acad. Sci. U.S.A.* **81**: 198–201.

Ochman, H., T.S. Whittam, D.A. Caugant, and R.K. Selander, 1983. *J. Gen. Microbiol.* **129**: 2115–2726.

Ochman, H. and A.C. Wilson, 1987. In *Escherichia coli and Salmonella typhimurium: Cellular and Molecular Biology*, edited by F.C. Neidhardt, J.L. Ingraham, K.B. Low, B. Magasanik, M. Schaechter, and H.E. Umbarger, pp. 1649–1654. American Society for Microbiology, Washington, D.C.

Ørskov, F. and I. Ørskov, 1983. *J. Infect. Dis.* **148**: 346–357.

Ørskov, F., I. Ørskov, D.J. Evans, Jr., R.B. Sack, D.A. Sack, and T. Wadstrom, 1976. *Med. Microbiol. Immunol.* **162**: 73–80.

Ørskov, I., F. Ørskov, B. Jann, and K. Jann, 1977. *Bacteriol. Rev.* **41**: 667–710.

Ørskov, F., T.S. Whittam, A. Cravioto, and I. Ørskov, 1990. *J. Infect. Dis.* **162**: 76–81.

Rayssiguier, C., D.S. Thaler, and M. Radman, 1989. *Nature* **342**: 396–401.

Sawyer, S., 1989. *Mol. Biol. Evol.* **6**: 526–538.

Schwesinger, M.D., 1977. *Bacteriol. Rev.* **41**: 872–902.

Selander, R.K., D.A. Caugant, H. Ochman, J.M. Musser, M.H. Gilmour, and T.S. Whittam, 1986. *Appl. Environ. Microbiol.* **51**: 873–884.

Selander, R.K., D.A. Caugant, and T.S. Whittam, 1987. In *Escherichia coli and Salmonella typhimurium: Cellular and Molecular Biology*, edited by F.C. Neidhardt, J.L. Ingraham, K.B. Low, B. Magasanik, M. Schaechter, and H.E. Umbarger, pp. 1625–1648. American Society for Microbiology, Washington, D.C.

Selander, R.K. and B.R. Levin, 1980. *Science* **210**: 545–547.

Selander, R.K. and T.S. Whittam, 1983. In *Evolution of Genes and Proteins*, edited by R.K. Koehn and M. Nei, pp. 89–114. Sinauer Associates, Sunderland, Massachusetts.

Sharp, P.M. and W.-H. Li, 1987. *Nucleic Acids Res.* **15**: 1281–1295.

Stoltzfus, A., J.F. Leslie, and R. Milkman, 1988. *Genetics* **120**: 345–358.

Tajima, F., 1989. *Genetics* **123**: 585–595.

Watterson, G.A., 1978. *Genetics* **88**: 405–417.

Weinstock, G. 1987. In *Escherichia coli and Salmonella typhimurium: Cellular and Molecular Biology*, edited by F.C. Neidhardt, J.L. Ingraham, K.B. Low, B. Magasanik, M. Schaechter, and H.E. Umbarger, pp. 1034–1043. American Society for Microbiology, Washington, D.C.

Whittam, T.S., H. Ochman, and R.K. Selander, 1983a. *Proc. Natl. Acad. Sci. U.S.A.* **80**: 1751–1755.

Whittam, T.S., H. Ochman, and R.K. Selander, 1983b. *Mol. Biol. Evol.* **1**: 67–83.

Whittam, T.S. and R.A. Wilson, 1988. *Infect. Immun.* **56**: 2458–2466.

Young, J.P.W., 1989. In *Genetics of Bacterial Diversity*, edited by D.A. Hopwood and K.F. Chater, pp. 417–438. Academic Press, New York.

Subject Index